3rd International Conference on Materials Engineering and Applications (ICMEA 2020)

IOP Conference Series: Materials Science and Engineering
Volume 894

Ho Chi Minh City, Vietnam
6 - 8 January 2020

ISBN: 978-1-7138-2691-0
ISSN: 1757-8981

Printed from e-media with permission by:

Curran Associates, Inc.
57 Morehouse Lane
Red Hook, NY 12571

Some format issues inherent in the e-media version may also appear in this print version.

Additional copies of this publication are available from:

Curran Associates, Inc.
57 Morehouse Lane
Red Hook, NY 12571 USA
Phone: 845-758-0400
Fax: 845-758-2633
Email: curran@proceedings.com
Web: www.proceedings.com

TABLE OF CONTENTS

MANUFACTURING AND PROCESSING

MATERIAL CHARACTERIZATION

IOP Conf. Series: Materials Science and Engineering **894** (2020) 011001 doi:10.1088/1757-899X/894/1/011001

Preface

The Third International Conference on Materials Engineering and Applications (ICMEA 2020) was held from 6 to 8 January 2020 in Ho Chi Minh City, Vietnam. The conference provided a wide platform for the participants to communicate with each other and to share their research achievements. The distinguished keynote speakers, on the other hand, created a motivational academic atmosphere through the presentation of their outstanding research outputs.

So, on behalf of the organizing committee, we are pleased to present this volume, containing selected contributions to this conference, of authors based in the industry, research institutes and universities. You will find that a variety of subjects around materials is being covered in these proceedings, including material´s science, processing and properties, as well as measuring methods and applications. The selection of papers for this volume, we must recognize, was indeed a challenging endeavor for the about fifty specialized researchers that voluntarily took their time for the reviewing process for this IOP Conference proceedings volume.

While we appreciate the efforts of all the author that submitted papers for this conference as well as the key role of the reviewers, we express our gratitude to all the members of the supporting staff, for their enduring effort to make the conference process a smooth one, and to the proceedings editing team, whose time and talent created the present volume from scratches.

Prof. Frank Otremba
Federal Institute for Material Research and Testing, Germany

Conference Co-Chairs

Prof. Takashige Omatsu, Chiba University, Japan
Prof. Frank Otremba, Federal Institute for Material Research and Testing, Germany
Prof. Dorrik Stow Frse, Heriot-Watt University, Edinburgh, Scotland, UK

Program Co-Chairs

Prof. Ahmad Zuhairi Abdullah, Universiti Sains Malaysia, Malaysia
Prof. Anja Pfennig, HTW-Berlin University of Applied Sciences, Germany

Technical Program Chair

Prof. Vivian W. Y. Tam, Western Sydney University, Australia

Publicity Chair

Assoc. Prof. Khoa Le, Western Sydney University, Australia
Prof. Ki-Taek Lim , Kangwon National University , South Korea

Technical Committees

Prof. Yalin Dong, University of Akron, USA
Prof. Guy Le Lay, Aix-Marseille Universit 茅, France
Prof. Chul Hee Lee, Inha university, South Korea
Prof. Ashraf Maher Abdel Ghaffar,National Center for Radiation Research and Technology, Egypt
Prof. Shimaa Mohamed Elsaeed Hamouda, Egyptian Petroleum Research Institute, Cairo, Egypt
Prof. Aly Anis Hamouda, University of Stavanger, Norway
Prof. Md. Munsur Rahman, Bangladesh University of Engineering and Technology, Bangladesh
Prof. Yarramaneni Sridharbabu, BML Munjal University, India
Assoc. Prof. Xifeng Li, Shanghai Jiao Tong University, China
Assoc. Prof. Sha Shiong Ng, Universiti Sains Malaysia, Malaysia
Assoc. Prof. Dr. Juraidah Ahmad, Universiti Teknologi MARA, Malaysia
Assoc. Prof. Sushil Kumar Jain, Manipal University Jaipur, India
Assoc. Prof. Bisweswar Ghosh, The Petroleum Institute, UAE
Dr. Alain Kusmoko, University of Wollongong, Australia
Dr. Waqas Mahmood, Shanghai Jiao Tong University, China
Dr. Fariza Mohamad, Universiti Tun Hussein Onn Malaysia, Malaysia
Dr. Chokchai Yatongchai, Maejo University, Thailand
Dr. Mohammadreza Heydartaemeh, Shahrood University of Technology
Dr. Nasrin Khalilzadeh, Nuclear Science and Technology Research Institute
Dr. Hasan M. A. ALBEGMPRLI, Northern Technical University
Dr. Yuhe Wang, Texas A&M University at Qatar, Qatar
Dr. Tarek M. Aboul-Fotouh, Al-Azhar University, Cairo, Egypt
Dr. Thitiphan Chimsook, Maejo University, Thailand
Dr. Raja Sekhar Dondapati, Lovely Professional University, India
Dr. Abdollah Esmaeili, National Iranian South Oil Company (NISOC), Iran Dr. Forough Ameli, Iran University of Science and Technology (IUST), Iran

IOP Conf. Series: Materials Science and Engineering **894** (2020) 011003 doi:10.1088/1757-899X/894/1/011003

First Name	Last Name
Anja	Pfennig
Son	Ngoc Truong
Lan Ching	Sim
Jayant	Jain
Thanat	Sitthisantikul
Sirayu Chanpee	Chanpee
Panupong Verasarut	Verasarut
Karl Vincent Alvarez	Alvarez
Chun-Ping	Lin
Bolin	ZHANG
Ruth	Aquino
PISUTTI	DARARUTANA
Nidal	Alshwawreh
Ruth	Aquino
Moumita	Sharma
Kengo	Aizawa
Kunal Sandip	Garud
Tomoya	Kawabata

Institution	Email Address
HTW-Berlin University of Applied Sciences,Germany	anja.pfennig@htw-berlin.de
Ho Chi Minh City University of Technology and Education, Vietnam	sontn@hcmute.edu.vn
Universiti Tunku Abdul Rahman,Malaysia	simcl@utar.edu.my
Indian Institute of Technology Delhi, India	jayantj@iitd.ac.in
King Mongkut's institute of technology ladkrabang, Thailand	stthanat@gmail.com
King Mongkut's institute of technology ladkrabang, Thailand	59110068@kmitl.ac.th
King Mongkut's institute of technology ladkrabang, Thailand	panupong331@gmail.com
University of the Philippines Diliman , Philippines	knalvarez@up.edu.ph
Asia University , Taiwan	chunping927@gmail.com
Nanjing Metro Group Co., Ltd , China	1120319381@qq.com
Mapua University , Philippines	rraquino@mapua.edu.ph
Retired Army Officer , Thailand	kritwonin@gmail.com
German Jordanian University, Jordan	nidal.alshwawreh@gju.edu.jo
Mapua University , Philippines	rraquino@mapua.edu.ph
Jadavpur University,India	moumitasharma21@gmail.com
Keio University, Japan	katsuranow@gmail.com
Dong-A University, Republic of Korea	kunalgarud1@gmail.com
The University of Tokyo, Japan	kawabata@fract.t.u-tokyo.ac.jp

IOP Conf. Series: Materials Science and Engineering **894** (2020) 011004 doi:10.1088/1757-899X/894/1/011004

Peer review statement

All papers published in this volume of *IOP Conference Series: Materials Science and Engineering* have been peer reviewed through processes administered by the proceedings Editors. Reviews were conducted by expert referees to the professional and scientific standards expected of a proceedings journal published by IOP Publishing.

IOP Conf. Series: Materials Science and Engineering **894** (2020) 012001 doi:10.1088/1757-899X/894/1/012001

Electrical Performance Comparison between Conventional Pi Shaped and Linear Shaped Thermoelectric Generators

Kunal Sandip Garud[1, a], Mahesh Suresh Patil[1, b], Jae-Hyeong Seo[2, c], You-Ma Bang[3, d], Chong-Pyo Cho[3, e] and Moo-Yeon Lee[2, f] *

[1]Graduate School of Mechanical Engineering, Dong-A University, 37, Nakdong-daero 550beon-gil, Saha-gu, Busan 49315, Republic of Korea
[2]Department of Mechanical Engineering, Dong-A University, 37, Nakdong-daero 550beon-gil, Saha-gu, Busan 49315, Republic of Korea
[3]Korea Institute of Energy Research,152 Gajeong-ro, Yuseong-gu, Daejeon 34129, Republic of Korea

[a]kunalgarud1@gmail.com, [b]msp692@gmail.com, [c]cheonchw@naver.com, [d]ymbang@kier.re.kr, [e]cnsever@kier.re.kr, [f] *mylee@dau.ac.kr

Abstract. The main objective of the present work is to show the superiority in performance of linear thermoelectric generator (TEG) over the conventional Pi shaped thermoelectric generator. Linear thermoelectric generator with various leg lengths as well as different lengths of p-type and n-type semiconductors are modeled. Power and efficiency of the linear thermoelectric generator with various leg lengths and different lengths of p and n legs are compared at the various hot side temperatures numerically using ANSYS 19.1 commercial software. Numerically predicted power of the linear thermoelectric generator is found within 5% error with corresponding theoretical power values. Results presented that, linear thermoelectric generator with a ratio of p leg length to total leg length equal to 0.56 showed higher power output and higher efficiency for all leg lengths as well as all hot side temperatures. Maximum power and maximum efficiency for linear thermoelectric generator with 0.56 length ratio are found 6% higher than maximum power and maximum efficiency of conventional Pi shaped thermoelectric generator with 0.5 length ratio.

1. Introduction

Patil et al. conducted the comparion of segmented and non-segmented thermoelectric generators for symmetrical and assymetrical leg geometry for various geometrical and operating parameters [1]. Haider et al. have shown that, variation in shape parameter has a significant effect on power output and efficiency [2]. Thermoelectric as well as thermomechanical performances of annular thermoelectric generator were investigated numerically by Fan et al. [3]. Segmented thermoelectric generator composed with number of semiconductor materials showed higher performance than conventional thermoelectric generator [4]. The main objective of the present work is to explore the alternative way in order to enhance the performance of the conventional thermoelectric generator. In order to fulfill the objective, linear shaped thermoelectric generator with different lengths of p and n leg were analyzed numerically to compare the power output and efficiency with each other as well as conventional Pi shaped thermoelectric generator.

2. Numerical Analysis Methodology

IOP Conf. Series: Materials Science and Engineering **894** (2020) 012001 doi:10.1088/1757-899X/894/1/012001

The physical domain of linear thermoelectric generator considered for the numerical is shown in Figure. 1.Similar to conventional Pi shape thermoelectric generator, linear thermoelectric generator also consists of p leg, n leg, top hot plate and bottom two cold plates. However, in case of linear thermoelectric generator, length of p leg (L_p) and length of n leg (L_n) can be varied which is not possible in case of Pi shaped thermoelectric generator. L presents the total length which is the sum of length of p leg and length of n leg. Properties of various parts of thermoelectric generator materials which are used in the numerical analysis is shown in Table 1.

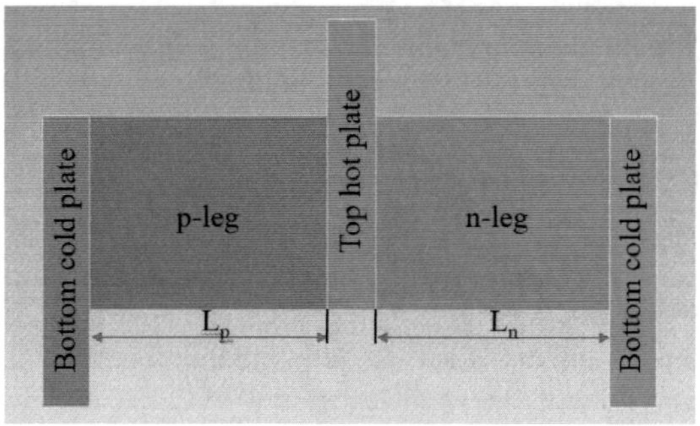

Figure. 1 Physical domain

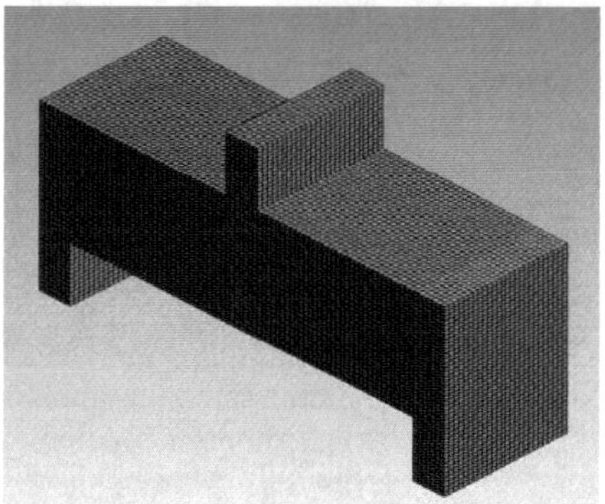
Figure. 2 Meshing in linear thermoelectric generator

Table 1 Properties of material used in the numerical analysis

Material	Thermal conductivity (W/m K)	Electrical resistivity (Ω m)	Seebeck coefficient (μV/K)
p leg	2.68-3.62	0.62×10^{-5}-0.81×10^{-5}	92-151.4
n leg	3.04-4.05	0.67×10^{-5}-1.17×10^{-5}	-226.5 to -162
Copper (hot and cold plates)	385	1.7×10^{-8}	-

Meshing for the considered physical domain of linear thermoelectric generator is presented in the Figure. 2. Sweep method with body sizing of 0.0003 m was used as the meshing configuration. Total of 67,516 mesh elements was generated with selected mesh configuration. Power of thermoelectric generator was simulated for five different mesh number and after mesh number of 67,516, power was converged within ±1%. Therefore, mesh number of 67,516 was selected as the final mesh for further detail analysis. Boundary conditions used in the numerical analysis is presented in the Table 2.

Table 2 Boundary conditions used in the numerical analysis

Parameter	Specification
Hot plate temperature (°C)	50, 100, 150, 200, 250, 300
Cold plate temperature (°C)	25
High potential voltage (V)	Optimum voltage calculated by Eq. (2)
Low potential voltage (V)	0

To analyze the electrical performances of the linear thermoelectric generator with length ratio of legs of the thermoelectric module, length ratio of legs was defined as Eq. 1.

$$Length\ ratio = \frac{L_p}{L} \tag{1}$$

Where, L_p is p leg length and L is total length which is equal to $L_p + L_n$

Eqs. (2) to (4) are derived from reference [2]. Optimum voltage is the voltage at which power and efficiency are maximum.

$$V = \alpha(T_1 - T_2) \tag{2}$$

Here, α is Seebeck coefficient. T_1 is hot side temperature and T_2 is cold side temperature.

Power can be calculated numerically using voltage and corresponding current values. Efficiency of thermoelectric generator is defined as a ratio of power output to the heat absorbed by TEGs.

$$P = VI \tag{3}$$

$$\eta = \frac{P}{Heat\ absorbed\ by\ thermoelectric\ generator} \tag{4}$$

Here, V is high potential voltage and I is current.

3. Analytical Analysis Methodology

To validate the numerically predicted results of the linear thermoelectric generator, the results are compared with theoretical results. Eqs. (5)-(8) are suggested by Jia et al. [5] to calculate the electrical performance parameters of thermoelectric generator.
Maximum power is presented as,

$$P_{max} = \frac{\alpha^2(T_1-T_2)^2}{4R_i} \tag{5}$$

Here, R_i is the internal resistance which is calculated as,

$$R_i = \frac{(\rho_p+\rho_n)L}{A} \tag{6}$$

Here, ρ_p and ρ_p are electrical resistivity of p and n legs, respectively. L is leg length and A is base area of leg.

Maximum efficiency is calculated as,

$$\eta_{max} = \frac{\alpha^2(T_1-T_2)R_e}{\alpha^2T_1(R_i+R_e)+K_i(R_i+R_e)^2-0.5\alpha^2(T_1-T_2)R_i} \tag{7}$$

Where, R_e is external resistance which is calculated as,

$$R_e = \sqrt{{R_i}^2 + \frac{\alpha^2R_i(T_1+T_2)}{2K_i}} \tag{8}$$

K_i is internal conductivity of thermoelectric generator.

4. Results

4.1. Validation

Maximum power values were predicted with respect to various hot side temperatures for the linear thermoelectric generator having leg lengths of 1 mm, 2 mm and 3 mm. For each leg length, length ratio is 0.5 means both p-type and n-type legs have equal lengths in each case. Linear tehrmoeelctric generator with same length of p leg and n leg is considered as the Pi shaped tehrmoelectric generator. Maximum power predicted using thermal electric module of ANSYS 19.1 as well as analytically calculated for each case. Comparison of numerically predicted and corresponding analytical results for the various leg lengths with length ratio of 0.5 is shown in Figure. 3. Error between the numerical and analytical results were more for the leg length of 1 mm however, it is below 5% over the entire temperature range. Errors between the numerical and analytical results for the leg length of 2 mm is within 2% and for leg length of 3 mm, error is within 2.5%. Numerically predicted results showed acceptable results with corresponding analytical results for the various leg lengths as well as temperatures.

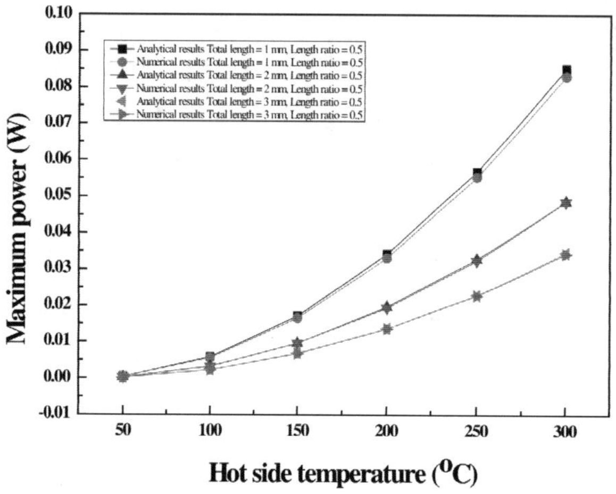

Figure. 3 Comparison of numerically predicted and corresponding analytical results for the various leg lengths with length ratio of 0.5

4.2. Effect of Length Ratio

Maximum power of linear thermoelectric generator for the leg length of 1 mm, 2 mm and 3 mm, and various length ratio ranging from 0.1 to 0.9 were predicted at the temperature of 300°C. Maximum power variation for various leg lengths and length ratio at the temperature of 300°C is shown in Figure. 4 (a). In case of conventional Pi-shape thermoelectric generator, maximum power°Ccurred for the equal length of p and n legs. However, in case of linear thermoelectric generator, it is found that, for all the leg lengths, maximum power is°Ccurred at the length ratio of 0.56 means length of p leg is higher than the n leg. Compared to maximum power at the length ratio of 0.5 which is similar to Pi shaped thermoelectric generator case, maximum power for the linear thermoelectric generator at the length ratio 0.56 are 2.3%, 4.4% and 5.7% higher for the leg lengths of 1 mm, 2 mm and 3 mm, respectively. Similarly, maximum efficiency of linear thermoelectric generator for various leg lengths and leg ratio is predicted at the temperature of 300°C which is presented in Figure. 4 (b). Maximum efficiency for the Pi shaped thermoelectric generator is°Ccurred at the 0.5 length ratio however, it is found at the length ratio of 0.56 for the linear thermoelectric generator. Maximum efficiency for the linear thermoelectric generator is 5.6% higher than the Pi shaped thermoelectric generator for all the leg lengths. From the maximum power and efficiency figures, it can be noticed that, as the leg length increased from 1 mm to 3 mm, maximum power has decreased whereas, maximum efficiency has increased. In present analysis, Pi shaped thermoelectric generator showed higher electrical resistance than linear thermoelectric generator which has resulted in lower power and efficiency of Pi shaped thermoelectric generator.

4.3. Effect of Total Leg Length

Variation of maximum power and maximum efficiency with respect to total leg lengths for various length ratio at temperature of 300°C are presented in Figures. 5 (a) and 5 (b), respectively. Total leg lengths varied from 1 mm to 3 mm with various length ratios of 0.1, 0.5, 0.56 and 0.9. From this result, it is found that, for all the leg lengths, maximum power is higher for length ratio of 0.56 than other length ratio. Here also, it is claimed that, linear thermoelectric generator with length ratio of 0.56 showed 4% higher maximum power than the maximum power of Pi shaped thermoelectric generator with length ratio of 0.5. Linear thermoelectric generator with length ratio of 0.56 showed 6% higher maximum efficiency than the maximum efficiency of Pi shaped thermoelectric generator with length ratio of 0.5

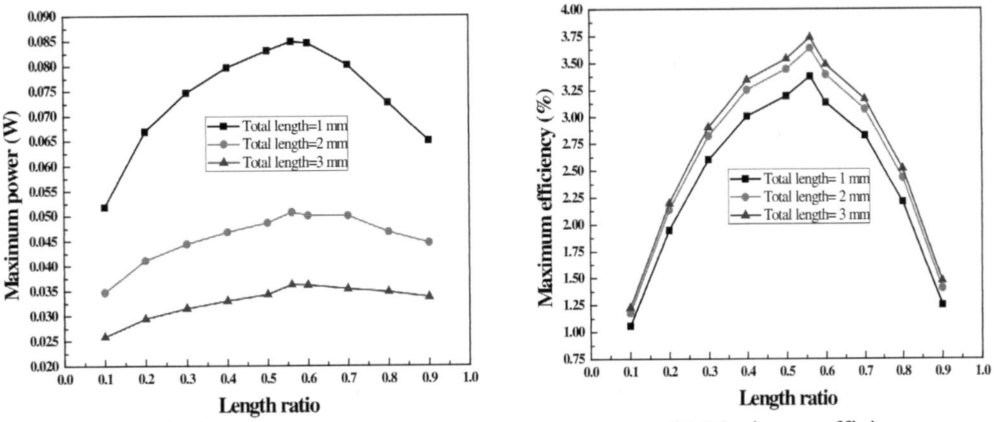

(a) Maximum power (b) Maximum efficiency

Figure. 4 Variation of (a) maximum power and (b) maximum efficiency for various leg lengths and length ratio at the temperature of 300°C

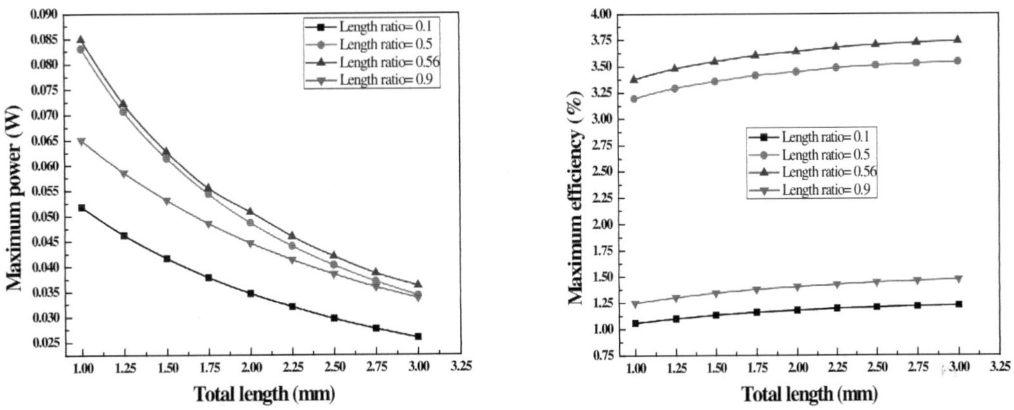

(a) Maximum power (b) Maximum efficiency

Figure. 5 Variation of (a) maximum power and (b) maximum efficiency with respect to total leg lengths for various length ratio at temperature of 300°C

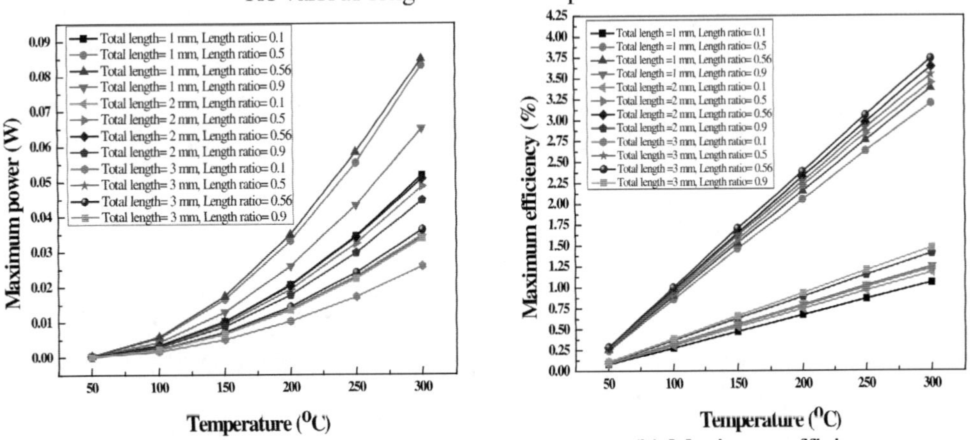

(a) Maximum power (b) Maximum efficiency

Figure. 6 Behavior of (a) maximum power and (b) maximum efficiency with respect hot side temperature for various leg lengths and length ratio

4.4. Effect of Temperature

Figure. 6 (a) and 6 (b) show the maximum power and maximum efficiency behavior with respect hot side temperature for various leg lengths and length ratio. For all the leg lengths as well as over the entire variation of temperature, length ratio of 0.56 which represents linear thermoelectric generator has maximum power and maximum efficiency values are higher than the corresponding values for the length ratio of 0.5 which represents Pi shaped thermoelectric generator for all the leg lengths.

5. Conclusions

Power and efficiency characteristics of linear thermoelectric generator were compared numerically with conventional Pi shaped thermoelectric generator. In case of Pi shaped thermoelectric generator p-type and n-type have same leg lengths however, leg lengths of p-type and n-type can be varied and it is possible to keep different lengths for both p and n legs. This characteristic of linear thermoelectric generator has resulted into higher power output and higher maximum efficiency. Linear thermoelectric generator was modelled with various leg lengths and length ratio. Effects of various leg lengths, length ratios and hot side temperatures were considered on the power and efficiency of linear thermoelectric generator. It is noticed that, length ratio of 0.56 for all the leg lengths and all the temperatures showed higher maximum power and maximum efficiency which were noticed for the length ratio of 0.5. Therefore, linear thermoelectric generator with different lengths of p and n legs was found better with approximately 6% higher power and efficiency than Pi shaped thermoelectric generator.

Acknowledgement

This work was supported by the Dong-A University research fund.

6. References

[1] Patil M.S, Seo J.H and Lee M.Y: International Journal of Air Conditioning and Refrigeration Vol. 26 (2018), p. 1-9 (185004).

[2] Ali, H., Sahin, A. Z., & Yilbas, B. S: Energy conversion and management Vol. 78 (2014), p. 634-640.

[3] Fan, S., & Gao, Y: Energy Vol. 150 (2018), p. 38-48.

[4] Ma, Q., Fang, H., & Zhang, M: Applied Thermal Engineering Vol. 127 (2017), p. 758-764.

[5] Jia, X., & Gao, Y: Applied Thermal Engineering, Vol. 78 (2015), 533-542.

IOP Conf. Series: Materials Science and Engineering **894** (2020) 012002 doi:10.1088/1757-899X/894/1/012002

Analysis of the Impact of Wire Resistance on Nano-scale Memristor Crossbar Array Implementing Perceptron Neural Network

Son Ngoc Truong

Faculty of Electrical and Electronics Engineering, Ho Chi Minh City University of Technology and Education, Vietnam

sontn@hcmute.edu.vn

Abstract. In this work, the impact of wire resistance in pure memristor crossbar array is mathematically analysed and verified by the circuit simulation. The memristor crossbar without CMOS device is utilized for application of character image recognition, in which wire resistance is presented. Memristor crossbar circuit is analysed separately with respect to wire resistance on vertical line and wire resistance on horizontal line. The result shows that wire resistance on vertical line can be eliminated because they can be self-compensated. The simulation result agrees with the analysis. The variation of output voltage caused by wire resistance less affect the recognition rate of memristor circuit. On the other hand, when wire resistance on horizontal line is assumed to be 2.5Ω. The output voltages are varied remarkably. Such variation of output voltage degrades the recognition rate of memristor crossbar circuit. The interesting phenomenon is also investigated. The column that is close to the first column has less variation of output voltage, whereas the one that is far from the first column has much variation of output voltage. The result can be used in improving the memristor crossbar architecture, which can tolerate the impact of wire resistance. For example, if we want to increase the size of crossbar, we should increase the number of rows, rather than the number of columns.

1. Introduction

Neuromorphic computing system which is inspired from human brain has been considered as an alternative solution to overcome the limitation of von Neumann computer architecture [1]. Neuromorphic refers to a hardware realization of artificial neural network mimicking the functionality of biological brain. For implementing neuromorphic systems, various research activities that are based on CPUs, GPUs, FPGAs, analog circuits, memory circuits, etc. have been carried out both in academia and industry [2]-[9]. Since memristor was experimentally demonstrated in 2008, it has become a potential nano-scale device for the design of neuromorphic computing system [10], [11]. Moreover, memristor can be formed as crossbar array and stacked layer by layer for high-density architecture [12]. More important, the charge-flux relationship of memristor is very close to the behavior of neural synapse [13]. Modeling neural synapse using memristor offers many advantages in term of power consumption and area occupation.

There are several ways to design neural synapse using memristor. The popular structures are the combination of memristor and CMOS devices [14]-[16]. Though such structures exhibit the excellent implementation of neural synapse, they seem to be limited since CMOS device can not be scaled down

IOP Conf. Series: Materials Science and Engineering **894** (2020) 012002 doi:10.1088/1757-899X/894/1/012002

more and more, and CMOS is power-hungry. Alternatively, pure memristor crossbar array without CMOS device is more efficient in term of the integration and power consumption [17]-[19]. Memristor crossbar array circuits gain more and more advantage in implementing neuromorphic computing system. Miao Hu et al proposed a crossbar of synaptic array that is composed of plus and minus crossbar array representing plus- and minus-polarity connection matrix [19] for implementation of neural network. The pure memristor crossbar array is very effective, compared to using memristor-CMOS hybrid circuits. To reduce the area, as well as the power consumption more, S. N. Truong proposed a new memristor crossbar array architecture which is composed of a single memristor array and a constant-term circuit. The proposed architecture shows that the power consumption is reduced by 48% and the area is reduced by 50% too [20]. Since then, pure memristor has demonstrated to be effective in applications such as speech recognition and image recognition [21], [22].

In pure memristor array, some amount of voltage drop can be caused by interconnect resistance, also known as wire resistance along row and column lines [23]-[24]. Such resistance degrades the performance of circuit seriously as the size of array is increased more and more [24]. In all of the previous work, the memristor is model using Verilog A, and the crossbar circuit is simulated using Cadence Spectre Simulation [19], [20]. For such approach, the impact of wire resistance is ignored, although it becomes serious problem in practical.

In this work, we analyse the impact of wire resistance to the performance of pure memristor crossbar array, which implements the perceptron neural network. The impact of wire resistance is mathematically analysed and then verified by the circuit simulation. The result emerges that the impact of wire resistance on vertical line can be compensated itself. Wire resistance on horizontal line degrades the performance of memristor crossbar remarkably. However, it is interesting that the variation caused by line resistance on horizontal line is proportional to the length of metal line. The analysis is more useful in improving the reliability of pure memristor crossbar array.

2. Analysis of the impact of wire resistance on crossbar-based synaptic array circuit

The improved memristor crossbar array for low power consumption and small area is shown in Figure 1. Here only one crossbar array is employed, associated with a constant term circuit for implementing negative and positive neural synapse [20].

In Figure 1, $g_{j,k}$ represents the memristor's conductance at the crossing point between the jth row and k[th] column. $V_{IN,j}$ is the input voltage applied to the jth row. $V_{C,k}$ is the column-line voltage on the kth column. In Figure 1, $V_{C,F}$ enters G_F that is inverting OP amp with the negative feedback resistor, R_{F1} to generate the reference voltage, V_F. V_F is propagated to all column lines from $V_{C,1}$ to $V_{C,n}$ through R_{F2}, as indicated in Figure 1. By applying Kirchhoff current law to the column line, $V_{C,F}$, we can calculate V_F with Eq. (1).

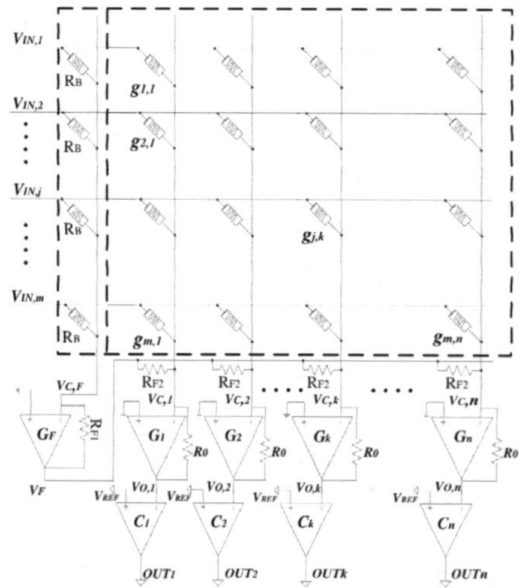

Figure 1. The improved memristor-based crossbar architecture with single memristor array and a constant-term circuit for implementing the negative and positive neural synapse [20].

$$V_F = -\sum_{j=1}^{m} \frac{R_{F1}}{R_B} V_{IN,j} \tag{1}$$

Similarly, we obtain the column voltage as follows.

$$V_{0,k} = -\left[\sum_{j=1}^{m} \left(R_0 \cdot g_{j,k} \cdot V_{IN,j}\right) + \frac{R_0}{R_{F2}} V_F\right] \tag{2}$$

If we choose $R_{F1}=R_{F2}$ and substitute V_F by Eq. (1) the following equation (3) can be obtained [20].

$$V_{0,k} = -\left[\sum_{j=1}^{m} \left(R_0 \cdot g_{j,k} - \frac{R_0}{R_B}\right) V_{IN,j}\right] \tag{3}$$

If $-\left(R_0 \cdot g_{j,k} - \frac{R_0}{R_B}\right)$ is defined as synaptic weight of the jth row and k[th] column, $w_{j,k}$, we can rewrite Eq. (3) with Eq. (4) .

$$V_{0,k} = \sum_{j=1}^{m} w_{j,k} V_{IN,j}$$

$$where\ w_{j,k} = R_0\left(\frac{1}{R_B} - g_{j,k}\right) = R_0\left(\frac{1}{R_B} - \frac{1}{M_{j,k}}\right) \tag{4}$$

In Eq. 4, the output voltage is a summation of all inputs, which are modified by their corresponding weights. Equation 4 figures out that each column has the functionality like a perception neuron, which has the output is a summation of all inputs multiplying with the corresponding synaptic weights, as shown in Figure 2(a) [25]. Figure 2(b) demonstrates that each column in memristor crossbar can implement a perceptron neuron, since the output of each column is consistent with the output of each perceptron neuron.

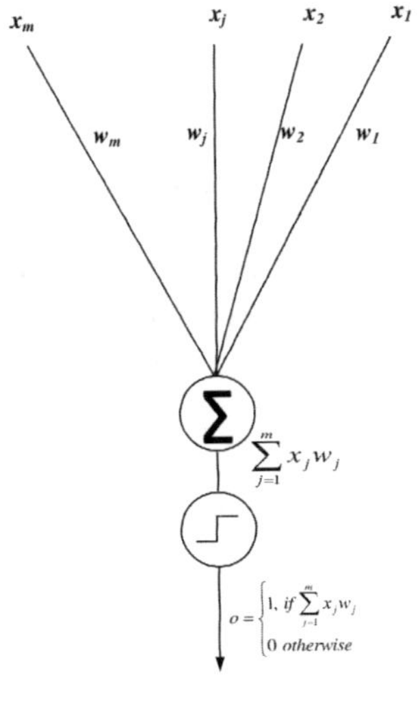

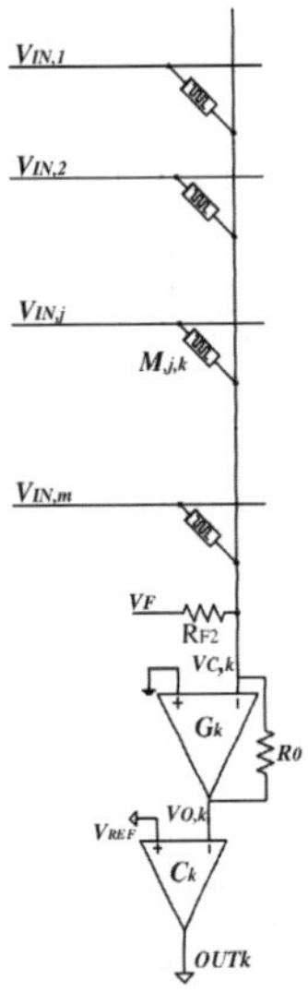

Figure 2. A perceptron neuron modeled by a sum of inputs with corresponding weights. (b) a perceptron neural modeled by one column in memrisitor crossbar.

In Figure 2(b), $M_{j,k}$ is programmable memristance value of the crossing point between the jth row and k[th] column. Resistor, R_B is a constant and R_B is chosen in a range of from low resistance state (LRS) to high resistance state (HRS). Then $w_{j,k}$ is programmed to be greater or less than R_B to generate negative or positive synaptic weight.

IOP Conf. Series: Materials Science and Engineering **894** (2020) 012002 doi:10.1088/1757-899X/894/1/012002

The output of each perceptron neuron is 0 or 1 decided by the threshold function [25]. The threshold function can be implemented by adding the comparator, C_k. as shown in Figure 1. By doing this, the neuron's output of kth column, OUT_k, can be activated or not. V_{REF} is the reference voltage for the comparators from C_1 to C_n. When $V_{O,k}$ is larger than V_{REF}, OUT_k becomes 1. If $V_{O,k}$ is smaller than V_{REF}, OUT_k is decided to be 0, as mathematically shown in Eq. (5).

$$OUT_k = \begin{cases} 1, \ if\ V_{0,k} \geq V_{REF} \\ 0, \ if\ V_{0,k} < V_{REF} \end{cases} \tag{5}$$

In memristor crossbar array, the voltage drops along column and row lines is inevitable [23], [24]. It becomes more serious as the size of array is increased more and more. Voltage drop is caused by the line resistance between the interconnections. The line resistance is modeled by the small-value resistor lying on the vertical lines and horizontal lines, as shown in Figure 3.

In this work, we analyse the impact of wire resistance by mathematically calculating the amount of variation of output voltage as wire resistance is taken into account. Here we define the variation of output voltage as a difference of voltage when wire resistance is assumed to be 0 Ω and that one when wire resistance is assumed to be 2.5 Ω

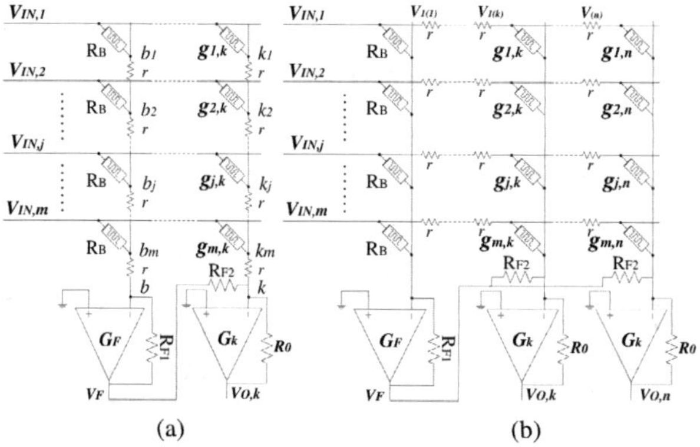

Figure 3. Wire resistance along horizontal and vertical lines are represented by small-value resistor, r connecting between two crossing points. (a) Wire resistances on horizontal lines are neglected. (b) Wire resistances on vertical lines are neglected.

We analyse the circuit separately with respects to wire resistance on vertical lines and wire resistance on horizontal lines, as shown in Figure 3(a) and (b), respectively. In Figure 3(a), we define V_{b1}, V_{b2} are the amount of voltage of node b_1 and b_2, which are on the first column. In common case, V_{bj} is the amount of voltage of node b_j on the first column. Similarly, V_{kj} is the amount of voltage of node k_j which is on the jth column. Applying Kirchhoff current laws for all nodes in Figure 2(a), V_F and $V_{O,k}$ can be estimated as follows:

$$\frac{-V_F}{R_{F1}} = \frac{V_{IN,m} - V_{bm}}{R_B} + \ldots + \frac{V_{IN,j} - V_{bj}}{R_B} + \ldots + \frac{V_{IN,1} - V_{b1}}{R_B}$$

$$V_F = -R_{F1}\left(\sum_{j=1}^{m} \frac{V_{IN,j}}{R_B} - \sum_{j=1}^{m} \frac{V_{bj}}{R_B}\right) \tag{6}$$

$$\frac{-V_{0,k}}{R_0} = \left(V_{IN,m} - V_{km}\right)g_{m,k} + \ldots + \left(V_{IN,j} - V_{kj}\right)g_{j,k} + \ldots$$

$$+ \left(V_{IN,1} - V_{k1}\right)g_{1,k} + \frac{V_F}{R_{F2}}$$

$$V_{0,k} = -R_0\left(\sum_{j=1}^{m} V_{IN,j}g_{j,k} - \sum_{j=1}^{m} V_{kj}g_{j,k} + \frac{V_F}{R_{F2}}\right) \tag{7}$$

If we assume that $R_{F1}=R_{F2}$, Eq. 7 can be simplified as follows:

$$V_{O,k} = -\left[\sum_{j=1}^{m}\left(R_0 \cdot g_{j,k} - \frac{R_0}{R_B}\right)V_{IN,j} - \sum_{j=1}^{m} R_0 V_{kj} g_{j,k} + \sum_{j=1}^{m} R_0 \frac{V_{bj}}{R_B}\right] \quad (8)$$

By comparing Eq. (8) and Eq. (4), we can derive the variation of voltage, ΔV caused by wire resistance on vertical lines as follow.

$$\Delta V = -\sum_{j=1}^{m} R_0 \frac{V_{kj}}{M_{j,k}} + \sum_{j=1}^{m} R_0 \frac{V_{bj}}{R_B} \quad (9)$$

$M_{j,k}$ is memristance which is programmed according to the synaptic weight using Eq. 4. For example, if the synaptic weigh, $w_{j,k}$ is negative, $M_{j,k}$ is less than R_B, and if $w_{j,k}$ is positive, $M_{j,k}$ is greater than R_B. Because the synaptic weight, $w_{j,k}$ obtained from the training process can have either negative or positive values, the variation of voltage, ΔV in Eq. (9) can be very small due to the compensation of the negative term of $-\sum_{j=1}^{m} R_0 \frac{V_{kj}}{M_{j,k}}$ and the positive term of $\sum_{j=1}^{m} R_0 \frac{V_{bj}}{R_B}$

Under this assumption, the variation of voltage, ΔV caused by wire resistance on vertical line can be compensated itself, and less affect to the output voltage.

Next, wire resistance on vertical lines is ignored whereas that one on horizontal lines is considered as shown in Figure 3(b). The input voltage for each column becomes different since the voltage is dropped on the small-value resistors which they pass through. In Figure 3(b), $V_{1(k)}$ is the amount of voltage on the resistor, which is on the first row and between the kth and $(k-1)^{th}$ column, the amount of voltage applied to first row of the kth column is significantly degraded and can be calculated as following equation.

$$V_{IN,1(k)} = V_{IN,1} - \sum_{i=1}^{k} V_{1(i)} \quad (10)$$

Here $V_{IN,1(k)}$ is the amount of voltage applied to the first row of the kth column. Then the output voltage of kth column, $V_{O,k}$ becomes

$$V_{O,k} = -\left[\sum_{j=1}^{m}\left(R_0 \cdot g_{j,k} V_{IN,j(k)} - \frac{R_0}{R_B} V_{IN,j}\right)\right] \quad (11)$$

By comparing Eq. (11) and Eq. (3), we obtain the variation of voltage, ΔV_k, caused by wire resistance on horizontal lines

$$\Delta V_k = \sum_{j=1}^{m} R_0 g_{j,k} V_{IN,j} - \sum_{j=1}^{m} R_0 g_{j,k} V_{IN,j(k)} \quad (12)$$

Here, $V_{IN,j(k)}$ can be calculated using Eq. (10), then we obtain ΔV_k, as shown in Eq. (13)

$$\Delta V_k = \sum_{j=1}^{m}\left(R_0 g_{j,k} \sum_{i=1}^{k} V_{j(i)}\right) \quad (13)$$

In Eq. (13), the term of $\sum_{i=1}^{k} V_{IN,j(i)}$ is the total voltage that drops on k resistors from the first column to the kth column of the jth row. Eq. (13) emerges that the amount of voltage applied to the kth column is degraded proportionally to distance from that column to the first column. Since the applied voltage is decreased, the output voltage obtained from the output of inverting OP amp can be increased. From this mathematical analysis, we can claim that the column that is close to the first column has less variation of voltage, and the column that is far from the first column has much variation of voltage. More interesting, two adjacent columns can have nearly same variation of voltage caused by wire resistance. The analysis result is very useful for design of memristor crossbar circuit, which can tolerate the effect of wire resistance.

3. Simulation and result

The SPECTRE circuit simulation is performed to verify the mathematical analysis of the impact of wire resistance. The memristor crossbar array circuit in Figure 1 is used for the application of

IOP Conf. Series: Materials Science and Engineering **894** (2020) 012002 doi:10.1088/1757-899X/894/1/012002

character image recognition. Figure 4 shows 8x8 character images used in this simulation. Each character image is composed of 8x8 pixels. To recognize 26 character images, the memristor crossbar is composed of 27 columns. The first column connects to all inputs through R_B to generate the negative voltage as mentioned in the previous section. And 26 columns are corresponding to 26 perceptron neurons for recognizing 26 character images from 'A' to 'Z'. The 64 rows correspond to 64 input voltages obtained from 64 pixels of character image to be recognized

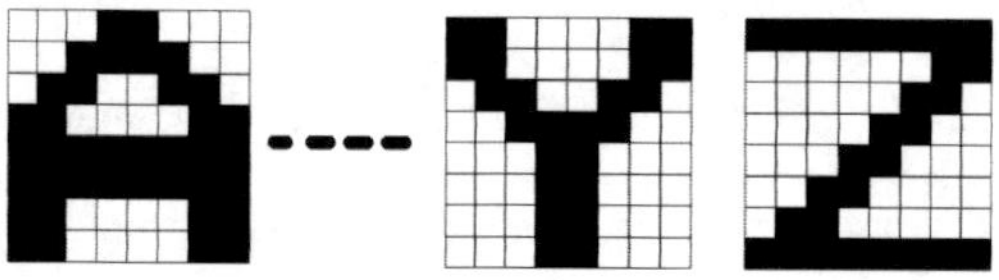

Figure 4. The 8x8 pixels images of characters used to test the memristor array circuit

In this simulation, we train 26 neurons to recognize 26 character images using chip-in-the-loop algorithm [20]. In testing phase, we apply the character image 'Z' to the crossbar and observer the output voltage of all columns when wire resistance on vertical line is 0 Ω and 2.5 Ω, respectively, wire resistance of horizontal line is ignored. The simulation result is shown in Figure 5. In Figure 5(a), when wire resistance is 0 Ω, the output voltage of the last column is close to 1V with respect to the input character image of 'Z', others are close to 0V, represented by square symbol in Figure 5(a). When wire resistance is 2.5 Ω, the change of output voltage can be observed, as represented by round symbol in Figure 5(a). However, the variation of voltage is not too large. If the threshold voltage is chosen to be 0.5 V, the crossbar can recognize the character 'Z' well. This exhibits that wire resistance on vertical line less affect to the performance of memristor crossbar circuit. The simulation result is consistent with the mathematical analysis in Eq. (9).

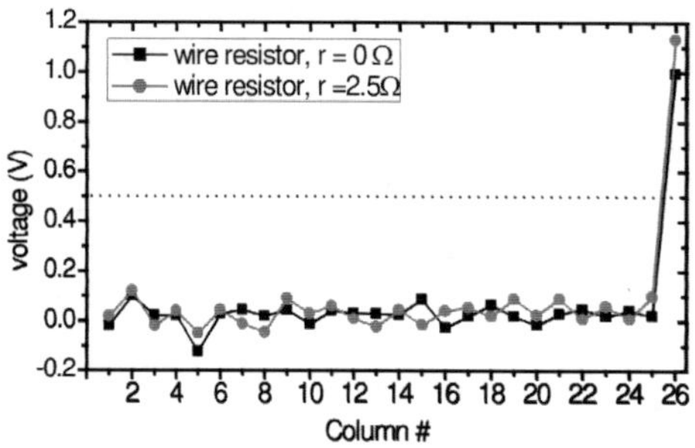

(a)

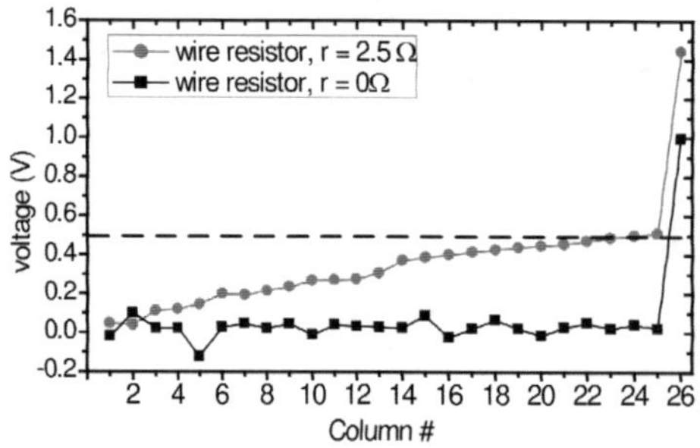

(b)

Figure 5. The output voltage of 26 columns with respect to the input character image 'Z'. (a) wire resistance on vertical lines are 0 Ω and 2.5 Ω, wire resistance on horizontal lines is ignored. (b) wire resistance on horizontal lines is 0, 2.5 Ω, respectively, whereas wire resistance on vertical lines is ignored.

Figure 5(b) shows the simulation result of memristor crossbar circuit for recognition of 26 character images, in which wire resistance on vertical line are ignored and wire resistance on horizontal line is assumed to be 0 Ω and 2.5 Ω, respectively. In Figure 5(b), we apply the character image of 'Z' to the input and obtain the column voltages. As the result, when wire resistance is assumed to be 2.5Ω, the output voltages of columns tend to increase, as represented by round symbol. Since the voltage drops on wire resistors, the actual amount of voltage applied to the last column is decreased remarkably. The output voltage of the last column is increased, as shown in Figure 5(b). It also shows that the column that is close to the first column has less variation of voltage, compared to that one is far from the first column, as illustrated in Figure 5(b). The simulation result is consistent with the analysis mathematically reasoned in Eq. (13). In Figure 5(b), if the threshold is kept to be 0.5 V, the columns of 23rd, 24th, and 25th are activated when the input is 'Z', meaning the crossbar recognizes the input character images incorrectly. Wire resistance on horizontal line strongly affects the performance of crossbar array circuit that implements the perceptron neural network for character image recognition.

In summary, in the memristor crossbar array implementing the perceptron neural network, wire resistance on vertical line can be eliminated by compensating itself. By contrast, wire resistance one horizontal line strongly affects the performance of the memristor crossbar circuit. From the analysis and the simulation, if we want to increase the size of crossbar, we should increase the number of rows, rather than the number of columns

4. Conclusion

In this work, the impact of wire resistance in pure memristor crossbar array is mathematically analyzed and verified by the circuit simulation. The crossbar is designed to implement the perception neural network for character image recognition. Wire resistance on vertical lines can be compensated itself, and less affect to the performance of the crossbar array. Wire resistance on horizontal lines degrades the recognition rate of the crossbar array significantly. It is interesting that the column that is close to the first column has less variation of output voltage, compared to the one is far from the first column. The circuit simulation results agree with the mathematical analysis. This analysis is useful. The result can be used to design the memristor circuits which can tolerate the effect of wire resistance. For example, if we want to increase the size of crossbar, we should increase the number of rows, rather than the number of columns.

5. References

[1] Mead C 1990 *Proceedings of the IEEE* **78** 1629
[2] Morns I P and Dlay S S 1999 *IEEE Trans. Neural Netw.* **10** 951
[3] Lont J B and Guggenbühl 1992 W *IEEE Trans. Neural Netw.* **3**
[4] Mirsa J and Saha I 2010 *Neurocomputing* **74** 239
[5] Mohamad K, Mahmud M F O, Adnan F H, and Abdullah W F H *IEEE Symposium on Humanities, Science and Engineering Research* 133
[6] Himavathi S, Anitha D, and Muthuramalingam A 2007 *IEEE Trans. Neural Netw.* **18** 880
[7] Al-Kazzaz S A 2008 *International Conference on Information and Communication Technologies* 1
[8] Granado J M, Vega M A, Pérez R, Sánchez J M, and Gómez J A 2006 *IEEE International Conference on Electronics, Circuit and System* 934
[9] Shima T, Kimura T, Kamatani Y, Itakura T, Fujita Y, and Iida T 1992 *IEEE J. Solid-State Circuits* **27** 1868
[10] Chua L O 1971 *IEEE Trans. Circuit Theory* **18** 507
[11] Strukov D B, Sinder G S, Stewart D R, and Williams R S 2008 *Nature* **453** 80
[12] Kügeler C, Meier M, Rosezin R, Gilles S, and Waser R 2009 *Solid State Electron* **53** 1287
[13] Jo S H, Chang T, Ebong I, Bhadviya B. B, Mazumder P, and Lu W 2010 *Nano Letters* **10** 1297
[14] Wang H, Li H, and Pino R E 2012 *International Joint Conference on Neural Networks*, pp. 1
[15] Kim H, Sad P, Yang C, Roska T, and Chua L O 2012 *IEEE Trans. Circuit and System* **59** 148
[16] Adhikari S P, Yang C, Kim H, and Chua L O 2012 *Trans. Neural Network and Learning System* **23**, 1426
[17] Chen Y C, Li H, Zhang W, Pino R E 2012 *IEEE Trans. Nanotechnology* **11**

IOP Conf. Series: Materials Science and Engineering **894** (2020) 012002 doi:10.1088/1757-899X/894/1/012002

[18] Liang J and Wong H S P 2010 *IEEE Trans. Electron Device* **57** 2531
[19] Hu M, Li H, Wu Q, Rose G S, and Chen Y 2012 *International Joint Conference on Neural Networks* 1
[20] Truong S N and Min K S 2014 *Journal of Semiconductor Technology and Science* **14** 356
[21] Truong S N, Ham S J, and Min K S 2014 *Nanoscale Research Letters* **9** 1
[22] Truong S N and Min K S 2015 *IEEE Transactions on Nanotechnology* **14** 1104
[23] Linn E, Rosezin R, Kügeler C, and Waser R 2010 *Nature Materials* **9** 403
[24] Shin S H, Byeon S D, Song J S, Truong S N, Mo H S, Kim D J, and Min K S 2015 *Journal of Semiconductor Technology and Science* **15**
[25] Mitchell T M 1997 (McGraw-Hill Science)

IOP Conf. Series: Materials Science and Engineering **894** (2020) 012003 doi:10.1088/1757-899X/894/1/012003

Synthesis and Characterization of Dried Leaves Derived Carbon Quantum Dots and g-C_3N_4 Composite

L C Sim[1], **J M Khor**[2], **K H Leong**[2] **and P Saravanan**[3]

[1] Department of Chemical Engineering, Lee Kong Chian Faculty of Engineering and Science, Universiti Tunku Abdul Rahman, Jalan Sungai Long 9, Bandar Sungai Long 43000 Kajang, Selangor, Malaysia
[2] Department of Environmental Engineering, Faculty of Engineering and Green Technology, Universiti Tunku Abdul Rahman, Kampar, Perak, Malaysia
[3] Department of Environmental Science and Engineering, Indian Institute of Technology(ISM), Dhanbad, Dhanbad-826004, Jharkhand, India

simcl@utar.edu.my

Abstract. In this work, carbon quantum dots (CDs) was successfully synthesized by hydrothermal treatment using dried leaves as green precursor. Graphitic carbon nitride (g-C_3N_4) was combined with CDs to prepare CDs/g-C_3N_4 composites with three different weight percentage at 0.6, 0.8 and 1.0 wt%, respectively. The morphological structure, optical properties and chemical compositions of CDs and composites were characterized using various spectroscopic techniques. CDs solution portrayed a significant fluorescence property that bright blue-green fluorescence can be observed by naked eye under ultraviolet (UV) light irradiation. The highest fluorescence emission was recorded at 320 nm with the optimal excitation wavelength of 423 nm. Ultraviolet-visible diffuse reflectance spectroscopy (UV-DRS) results displayed red-shifted adsorption spectra of CDs/g-C_3N_4 composites from 500 nm to 800 nm. No upconversion photoluminescence (UCPL) was detected in CDs based on the photoluminescence (PL) study. The loading of CDs on g-C_3N_4 reduced the band gap from 2.7 eV to 2.59 eV.

1. Introduction

In the early of 21st century, carbon quantum dots (CQDs) were first discovered in a research for purification of fluorescent single-walled nanotubes (SWNTs) [1] whereby it was later on widely known as carbon dots (CDs). In general, CDs are semiconductor nanoparticles with the diameter size of less than 10 nm [2] that composed of graphitic carbon (sp^2-hybridization) and contributed to quantum confinement effect. With this discovery, CDs and graphene quantum dots (GQDs) which both have the similar structural properties were getting significant interest and attention in nanotechnology and scientific field. This is due to their distinct advantages in terms of properties such as low toxicity [3], tuneable and comparable photoluminescence (PL) [4] chemical stability [5], and a wide range selection of environmental friendly synthesis process [2]. In recent years, hydrothermal method is widely used to produce CDs because this method is low toxicity, controllable size, can be prepared in the large amount, ability to recycle waste and safe. The CDs were prepared through hydrothermal treatment by using different precursors such as orange juice, strawberry juice, grass, waste paper, and other chemicals. The hydrothermal treatment of CDs usually involves the use of biomaterials, waste and less hazardous precursor and thus it is more environmentally friendly.

IOP Conf. Series: Materials Science and Engineering **894** (2020) 012003 doi:10.1088/1757-899X/894/1/012003

Hydrothermal is the term that refers to any heterogeneous chemical reaction in the presence of an aqueous or non-aqueous solvent above room temperature and greater than atmospheric pressure in a closed system [6]. Hydrothermal method is chosen to produce high quantum yield CDs by using low level of oxygen heating. Moreover, the reaction times are reduced with a good control of the crystallization, crystal size, purity and morphology of the products. The g-C_3N_4 was widely used in photoluminescence applications such as photodegradation of Rhodamine B and methyl orange, activation of dioxygen and converting CO_2 to fuel. In the domain of photocatalytic application, g-C_3N_4 tend to produce lower quantum efficiency that is about 1 % at 420 nm due to its fast electron charge recombination and poor conductivity [7]. Instead of using titanium dioxide (TiO_2) that had been widely chosen as one of the promising semiconductor in heterogeneous photocatalysis, g-C_3N_4 dominates TiO_2 since it is more economical, high thermal and chemical stability and better visible absorption properties and a greener semiconductor selection without the use of metal. The low quantum efficiency of g-C_3N_4 in photocatalytic application can be solved by incorporating highly photoluminescent CDs into the g-C_3N_4 to form hybridized CDs/g-C_3N_4.

2. Materials and methods

2.1. Preparation of CDs solution

Dried Bucida (Buccida Buceras), Kelat Paya (Eugenia Oleina), Indian Rosewood (Dalbergia Latifolia) and Hop Tree (Arfeuillea Arborescens) leaves were obtained from Universiti Tunku Abdul Rahman, Kampar campus, Malaysia. These leaves were dried in the oven for 12 h at 120 °C. 5 g of the leaves were mixed with 100 mL of ultrapure water in a 150 mL Teflon-lined stainless steel autoclave for 12 h at 200 °C. The dark brown solution was centrifuged for 20 min at 5,000 rpm to separate the large particles from the solution.

2.2. Preparation of CDs/g-C_3N_4

Pure g-C_3N_4 was prepared by putting urea (Merck, 99%) into an alumina crucible with lid and heated in muffle furnace at 550 °C for 3 h. The yellowish product, g-C_3N_4 (0.1 g) was then mixed with 1 mg/mL of CDs solution and heated at 100 °C for 2 h in Teflon-lined stainless steel autoclave. The obtained product was centrifuged at 5000 rpm for 15 min and washed with deionized water for several times. It was then dried at 60 °C overnight. By following the same procedure, the composites of CDs/g-C_3N_4 with different amount of CDs (0.6, 0.8, 1.0 wt%) were prepared and denoted as CDs/g-C_3N_4-0.6, CDs/g-C_3N_4-0.8, and CDs/g-C_3N_4-1.0, respectively.

2.3. Characterization of CDs/g-C_3N_4

The morphological and elemental information were determined by using field emission scanning electron microscopy (FESEM, JEOL 6701-F) and energy dispersive X-ray spectroscopy (EDX). The photoluminescence (PL) spectra of CDs solution were acquired with a PL spectrophotometer (Perkin Elmer LS 55). The ultraviolet-visible diffuse reflectance spectra (UV-Vis DRS) was obtained using Shimadzu UV-2600 spectrophotometer equipped with integrating sphere attachment with $BaSO_4$ as a reference.

3. Results and discussion

3.1. FESEM and EDX

FESEM was carried out to identify the morphology and structure of CDs, g-C_3N_4 and CDs/g-C_3N_4 composites. Figure 1(a) shows the FESEM image of CDs which observed to have irregular shapes with the particles size distribution ranging from 1μm to 2.86 μm. FESEM image of g-C_3N_4 in figure 1(b) illustrate the shape of g-C_3N_4 to be irregular crumpled lamellar structure. In figure 1(c-e), it clearly shows that wrinkled structure of g-C_3N_4 provided a better support for the attachment of CDs at different concentration where higher concentration CDs/g-C_3N_4 composite exhibited a more packed and aggregated morphological structure. In order to identify the elemental composition of CDs/g-C_3N_4 composites, EDX was carried out to confirm the presence of carbon (C), nitrogen (N) and oxygen (O)

IOP Conf. Series: Materials Science and Engineering **894** (2020) 012003 doi:10.1088/1757-899X/894/1/012003

in CDs/g-C_3N_4-0.8 composite. The weight percentage of C, N and O were found to be 37.13, 57.45 and 5.42 %. The C and N elements originate from g-C_3N_4 while O element was contributed by the reaction between urea and oxygen during the preparation of g-C_3N_4.

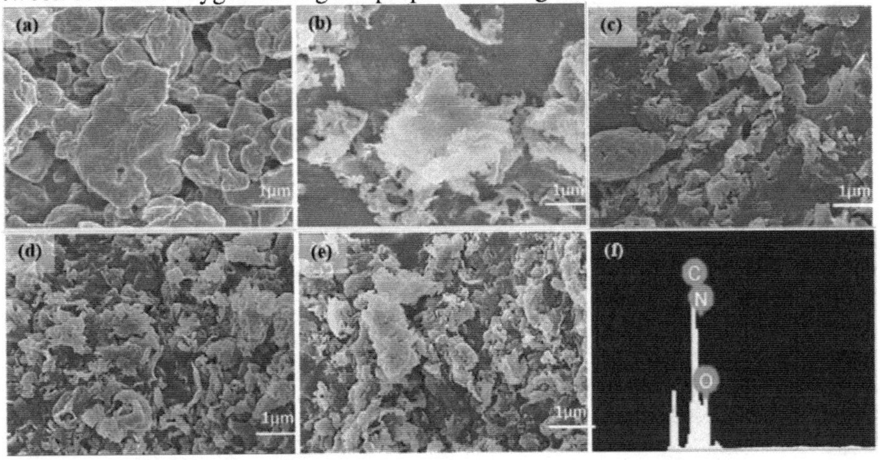

Figure 1. FESEM images of (a) CDs; (b) g-C_3N_4; (c) CDs/g-C_3N_4-0.6; (d) CDs/g-C_3N_4-0.8; (e) CDs/g-C_3N_4-1.0; and (f) EDX of CDs/g-C_3N_4-0.8.

3.2. Optical properties

Optical properties of CDs are considered one of the crucial factor that will affect the applications of CDs. The colour of CDs solution under daylight is brown as shown in the inset of figure 2. The CDs solution shows a bright blue-green fluorescence colour under the irradiation of ultraviolet (UV) light. The blue-green fluorescence of CDs that can be observed by naked eye visually confirmed its fluorescence characteristic that renders it a promising candidate in broad range of applications [8]. Figure 2 shows the UV-vis absorption spectrum of CDs solution in the range of 200 nm to 800 nm. There is only one absorption peaks at 303 nm which is attributed to n-π^* transition of C=O bonds in CDs [9].

Figure 3 illustrates the photoluminescence (PL) emission spectra behaviour across the excitation wavelength ranging from 300 nm to 500 nm. The excitation wavelength at 320 nm recorded the optimal emission wavelength at 423 nm. When the excitation wavelength increases, the emission spectrum are red-shifted to higher wavelengths with lower emission intensity. These observations are consistent with the previous reports [10, 11]. Although the origin of PL luminescence in CDs is not fully clarified, the excitation-dependent PL behaviour could be interrelated to the different size distribution of CDs, multiple distribution of emissive site on passivated CDs and photogenerated charges recombination at surface-confined defect states [12, 13, 14]. Besides, there is no emission peak beyond the excitation wavelength of 500 nm, showing the absence of upconversion photoluminescence (UCPL) properties in CDs. In order for UCPL to occur, multiphotons must be activated simultaneously by absorbing two or more photons at the same time which emission wavelength emitted is shorter than the excitation wavelength [15]. Previous study reported that most of the CDs did not possess upconversion fluorescence. The detected UCPL properties actually originated from normal fluorescence excited by leaking component from second diffraction in the monochromator of the fluorescence spectrophotometer [16].

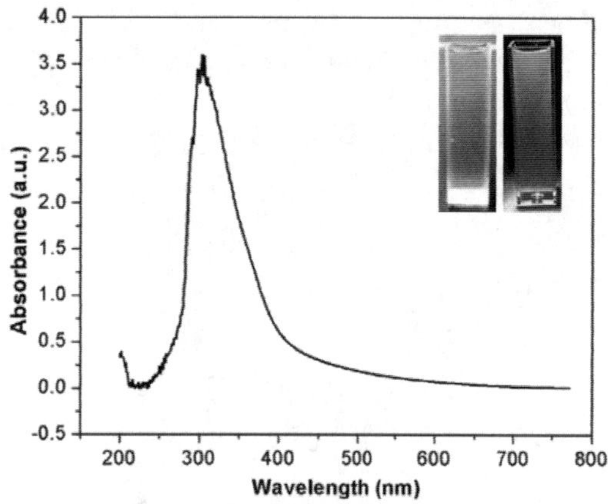

Figure 2. The UV-vis absorption spectrum of CDs solution. Inset shows the fluorescence of CDs solutions under daylight (left) and UV light (right).

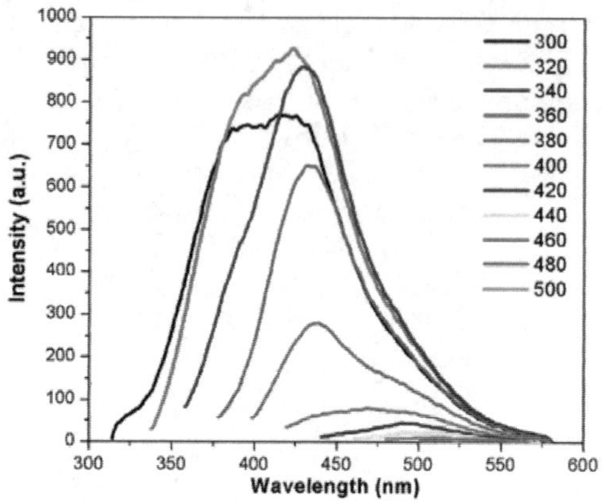

Figure 3. Photoluminescence emission wavelength of CDs solution at different excitation spectrum.

The study of ultraviolet-visible diffuse reflectance spectroscopy (UV-DRS) is shown in figure 4(a). The absorption spectra for all CDs/g-C_3N_4 composites is blue-shifted towards UV region before the cut-off wavelength at 450 nm but the absorption spectra start to red-shift towards visible region after 450 nm. The band gap of photocatalyst can be determined using Tauc plot where $(F(R)*hv)^{1/2}$ was plotted against photon energy, *hv* as shown in figure 4(b). The band gap of each photocatalysts is obtained from x-intercept by drawing a tangent line of each curve. The band gap for g-C_3N_4 is 2.70 eV while the band gap for CDs/g-C_3N_4-0.6, CDs/g-C_3N_4-0.8, and CDs/g-C_3N_4-1.0 composites is 2.68 eV, 2.65 eV and 2.59 eV, respectively. This results indicate that the increase of CDs weight percentage in the composite will lead to the reduction of band gap energy for improved visible light absorption. It is summarized that the light harvesting capability of photocatalysts follows the sequence of CDs/g-C_3N_4-1.0 > CDs/g-C_3N_4-0.8 > CDs/g-C_3N_4-0.6 > g-C_3N_4.

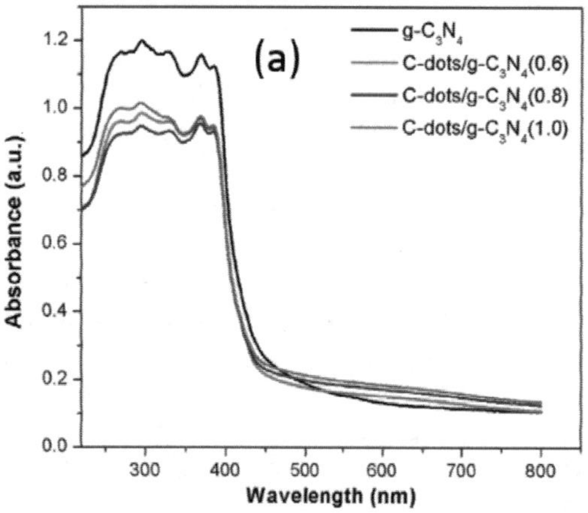

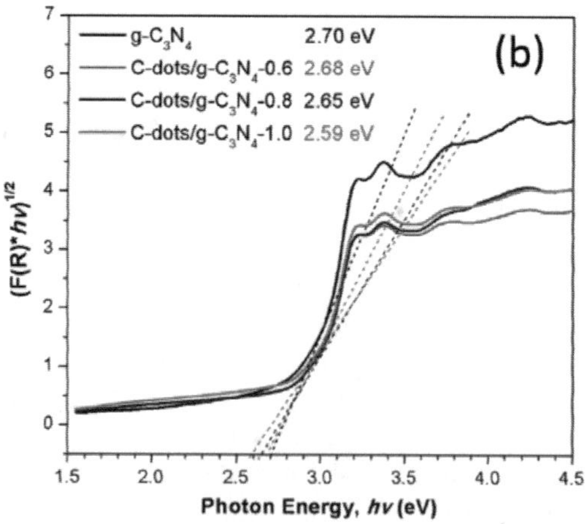

Figure 4. (a) UV-visible diffuse reflectance spectra of g-C_3N_4, CDs/g-C_3N_4-0.6, CDs/g-C_3N_4-0.8, and CDs/g-C_3N_4-1.0; (b) Tauc plots of g-C_3N_4, CDs/g-C_3N_4-0.6, CDs/g-C_3N_4-0.8, and CDs/g-C_3N_4-1.0.

4. Conclusions

CDs were successfully synthesized by using dried leaves as starting material through hydrothermal treatment. The PL emission spectra indicated the optimal emission wavelength of CDs at 423 nm upon excitation at 320 nm wavelength. The UCPL properties was not observed in all CDs/g-C_3N_4

composites. All CDs/g-C_3N_4 composites displayed the extension of light absorption range towards visible and NIR region from 500 nm to 800 nm. No upconversion photoluminescence (UCPL) was detected in CDs based on the photoluminescence (PL) study. The band gap energy was reduced from 2.7 eV to 2.59 eV after loading 1 wt% of CDs onto g-C_3N_4. Overall, the preparation of CDs/g-C_3N_4 composites are considered cost effective and environmentally friendly as no harmful chemicals had been used and dried leaves is a green precursor.

Acknowledgments

This work was supported by the Fundamental Research Grant Scheme (FRGS/1/2019/TK10/UTAR/02/5).

5. References

[1] Xu X Y, Ray R, Gu Y L, Ploehn H J, Raker L G K and Scrivens W A 2004 *J. Am. Chem. Soc.* **126** 12736
[2] Shen L M and Liu J 2016 *Talanta* **156-157** 245
[3] Sachdeva A and Packirisamy G 2015 *Analyst* **140** 4260
[4] Wang R, Lu K Q, Tang Z R and Xu Y J 2017 *J. Mater. Chem. A* **5** 3717
[5] Zhu S J, Meng Q N, Wang L, Zhang J H, Song Y B, Jin H, Zhang K, Sun H C, Wang H Y and Yang B 2013 *Angew. Chem.* **52** 1
[6] Ortiz-Landeros J, Gómez-Yáñez C, López-Juárez R, Dávalos-Velasco I and Pfeiffer H 2012 *J. Adv. Ceram.* **1** 204
[7] Sim L C, Tan W H, Leong K H, Bashir M J K, Saravanan P and Surib N A 2017 *Materials* **10** 28
[8] Guo Y M, Zhang L F, Cao F P and Leng Y M 2016 *Sci. Rep.* **6** 35795
[9] Niu W J, Li Y, Zhu R H, Shan D, Fan Y R and Zhang X J 2015 *Sensor Actuat. B: Chem.* **218** 229
[10] Martindale B C M., Hutton G A M, Caputo C A and Reisner E 2015 *J. Am. Chem. Soc.* **137** 6018
[11] Sim L C, Wong J L, Hak C H, Tai J Y, Leong K H and Saravanan P 2018 *Beilstein J. Nanotech.* **9** 353
[12] Ghosh S, Chizhik A M, Karedla N, Dekaliuk M O, Gregor I, Schuhmann H, Seibt M, Bodensiek K, Schaap I A T, Schulz O, Demchenko A P, Enderlein J and Chizhik A I 2014 *Nano Lett.* **14** 5656
[13] Sun Y P, Zhou B, Lin Y, Wang W, Fernando K A S, Pathak P, Meziani M J, Harruff B A, Wang X, Wang H F, Luo P G, Yang H, Kose M E, Chen B L, Veca L M and Xie S Y 2006 *J. Am. Chem. Soc.* **128** 7756
[14] Cao L, Mohammed J M, Sushant S and Sun Y P 2012 *Acc. Chem. Res.* **46** 171
[15] Wang Y F and Hu A G 2014 *J. Mater. Chem. C* **2** 6921
[16] Wen X M, Yu P, Toh Y R, Ma X Q and Tang J 2014 *ChemComm.* **50** 4703

IOP Conf. Series: Materials Science and Engineering **894** (2020) 012004 doi:10.1088/1757-899X/894/1/012004

Fabrication and Characterization of Polycaprolactone / Nanozeolite Membrane Films via Phase Inversion Technique

Ruth R Aquino[1,2], Marvin S Tolentino[1,3], MariaJenina S Foronda[4], Deseree A Mata[4] and Keille Peter B Susbilla[4]

[1]School of Chemical, Biological, and Materials Engineering and Sciences, Mapúa University, 658 Muralla St., Intramuros, Manila City 1002, Philippines
[2]Center for Nanomaterials and Membrane Research, Mapúa University, 658 Muralla St., Intramuros, Manila 1002, Philippines
[3]Industrial Technology Development Institute, Department of Science and Technology, DOST Compound, Gen. Santos Ave., Bicutan, Taguig City 1631, Philippines
[4] Senior High School, Mapúa University, 658 Muralla St., Intramuros, Manila City 1002, Philippines

rraquino@mapua.edu.ph, marvstolentino@yahoo.com

Abstract. This research focused on the fabrication and characterization of polycaprolactone (PCL)/nanozeolite(NZ) membrane films. PCL/NZ membrane blends were produced at varying amounts of NZ (0 wt. %, 0.5 wt. % and 1.0 wt. %).The membranes produced were characterized using a.) Scanning Electron Microscopy (SEM) in determining the surface morphologies, b.) UTM to evaluate the effect of NZ addition on the tensile strength of PCL, and c.)contact angle goniometer to investigate the effect of NZ on the wettability of PCL/NZ membrane blends. Upon characterization, the SEM result showed that increasing NZ concentration increased the presence of pores and pore-like structures on the PCL/NZ system. Also, the tensile strength of the PCL/NZ blend improved with the addition of NZ. Lastly, the contact angle measurements obtained showed enhanced membrane hydrophilicity because of the hydrophilic NZ present on the matrix. The incorporation of NZ on PCL generally improved its properties based the different characterizations conducted.

1. Introduction

Tremendous amounts of plastic materials are being produced and utilized each year due to numerous desirable properties being exhibit by this type of materials. Plastic materials are being used in many industries such electronics, construction, medicine and automotive because of their good performance throughout their service life and durable mechanical properties when exposed to weathering and harsh environments [1]. But accompanied by the increasing plastic material utilization is its increasing detrimental effect on the environment due to improper disposal of plastic waste. With this as a concern, consumers are shifting to the use of biodegradable polymers which addresses the disposal issues of plastic materials that usually ends in landfills [2].

Due to catastrophic environmental impact of plastic wastes, researchers focused on biodegradable plastics derived from renewable biomass which can be biologically degraded into components that are not harmful to the environment. In this regard, the production for biodegradable plastics increases due to its diverse applications [3]. One of the biodegradable plastics that has promising mechanical properties, and high potential for wide applications is polycaprolactone (PCL) [4]. But despite its good characteristics, considering that it is a biodegradable polymer, PCL still needs improvement in terms of its technical performance, especially when compared to several conventional polymers [2]. Due to this reason, enhancement of polymer properties can be done by fabricating nanocomposites through the addition of nanofillers[2,5].

Several studies concerning on the improvement of PCL properties were conducted by researchers. In particular, fabrication of nanofibrousnanocomposite of PCL using nanofillers were studied by other researchers as a potential material for wastewater treatment. In the study conducted by Irandoost et *al.*, (2019)[6], they fabricated PCL nanofibrousnanocomposites by incorporating nanoclay and nanozeolite (NZ) to improve its Pb^{2+} adsorption from the prepared aqueous solution. They found out that increasing clay minerals concentrations resulted to the increase in the Pb^{2+} adsorption by the PCL nanofibrousnanocomposite. Furthermore, improvement of heavy metal adsorption of electrospunnanofibers by nanoclay addition is evident in the study of Tolentino et *al.* (2019)[7]. In their study, incorporation of iron-modified NZ to the cellulose acetate matrix increased its adsorption capacity for the removal of Ni^{2+} ions from simulated wastewater.

Most researchers focus on heavy metal removal application of PCL reinforced with nanoclays through adsorption experiments. But the interaction of between PCL and nanoclays, such as NZ, as well as the mechanical properties should also be taken into account. The aim of this study was to produce PCL/NZ reinforced membrane films. Specifically, the objectives of this study were the following: a.) fabricate PCL/NZ membrane films at different NZ loadings through hand-casting method; b.) characterize the surface morphologies and cross-sections of the membrane films produced through Scanning Electron Microscopy (SEM); c.) determine the mechanical properties of the membranes by means of Universal Testing Machine (UTM); and lastly, d.) investigate membrane-water interaction through the use of contact angle goniometer.

This research can be a fundamental study for PCL/NZnanocomposite and can serve as a basis in determining the potential application for similar component systems.

2. Methodology

Different membrane blends were fabricated using PCL and NZ. PCL was first added to the solvent n-methyl-2-pyrollidone (NMP) with constant stirring until a homogeneous mixture was produced. Then, varying amounts of nanozeolite were added to produce different PCL/NZ blends (0 wt. % or pure PCL, 0.5 wt. % NZ and 1.0 wt. % NZ). The resulting mixtures continuously stirred and heated to 100°C until homogenization occurred. The mixtures were cooled down to room temperature to remove the bubbles brought about by the constant stirring. Hand-casting process was done by pouring each resulting mixture on the glass plate, and was spread evenly using a casting knife with 30 μm thickness. The films produced were immersed in the coagulation bath containing deionized water for 20 minutes to facilitate membrane formation. The membranes produced were set aside and air-dried for 24 hours. Characterization of the different membrane blends produced were conducted using the following techniques: a.) SEM (Hitachi S-3400N) for surface morphology and cross-section analyses; b.) UTM (Shimadzu UTM Model AGS-IKNX) for mechanical properties determination; and lastly, c.)contact angle goniometer (Theta Optical Tensiometer) for wettability assessment. Statistical analysis (ANOVA and post hoc test) of the contact angle measurements were also done to further verify the significance of NZ addition on PCL in terms of wettability.

3. Results and discussion

3.1. Surface morphology and cross-section analysis

Based on the results of SEM imaging (Figure 1), considering both the surface morphology and the cross-sectional image obtained, the appearance of pores was observed upon the addition of NZ on PCL. Moreover, the presence of pores or pore-like structures was evident with increasing NZ concentration (comparing Figure 1.a and Figure 1.b). Going in to detail, this phenomenon was due to the membrane fabrication technique that was used, which was phase inversion or phase separation process. Initially, PCL is a hydrophobic material due to its long hydrocarbon chains. Meanwhile, NZ is hydrophilic primarily because of the hydroxyl groups (-OH) present on its structure. During the phase inversion process, wherein the casted mixture was placed in the coagulation bath, the PCL/NZ mixture tend to attract water molecules on its system due to hydrophilic NZ that was present. In addition to this, the solvent for the PCL/NZ mixture used was NMP, which has a good miscibility with water. These factors can be correlated to the phase inversion exchange rate between the solvent, NMP, and water. This resulted to a higher possibility of having PCL/NZ (polymer) – water interaction. Once the membranes produced were removed from the coagulation bath and air-dried, the water molecules that underwent phase inversion with the solvent in the PCL/NZ system evaporated which produced to void spaces in the membrane system, and were regarded as pores[8].

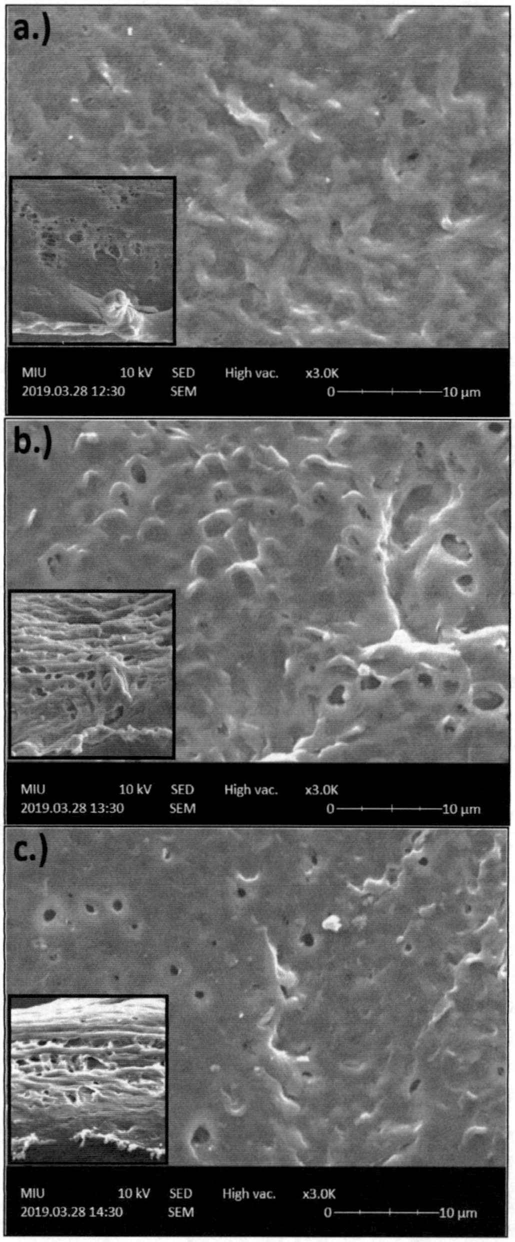

Figure 1.SEM images of the different membrane blends a.)pure PCL, b.) 0.5 wt. % NZ, and c.) 1.0 wt. % NZ at 3000x magnification (surface morphology and cross-section).

3.2. Determination of mechanical properties

To investigate the mechanical properties of the different membrane blends produced, specifically the tensile strength, a UTM was utilized. Based on the results obtained which was shown in Figure 2, the incorporation of NZ on PCL resulted to the increase on the membrane's tensile strength. Improved membrane tensile strength can be attributed to the strong interfacial bonding present between NZ particles and PCL polymeric chains [8].

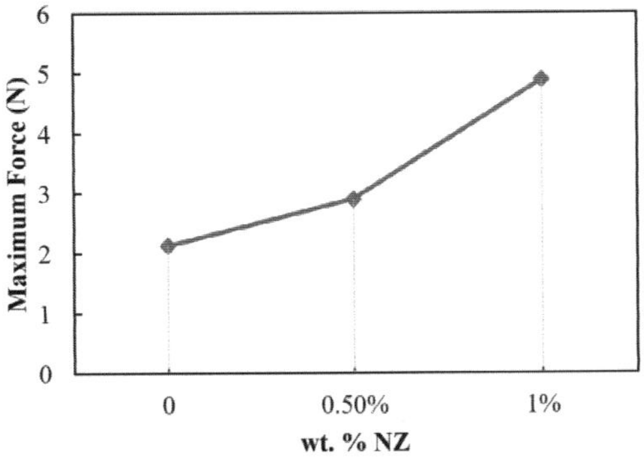

Figure 2. The tensile strength of the different membrane blends from the UTM.

3.3. Assessment of the membranes' wettability

The water contact angle for each membrane blend was measured using the goniometer. The results (Figure 3) showed that the addition of NZ particles on the PCL polymer matrix caused a significant decrease in terms of contact angle measurement. This decrease in contact angle measurement means increase in the membrane's hydrophilicity. Initially, pure PCL is hydrophobic in nature but upon the addition of the hydrophilic NZ, it can be seen that PCL/NZ membranes tend to become hydrophilic. NZ particles significantly increased the surface hydrophilicity of the PCL, which can be a great membrane property for wastewater treatment application since this would mean increase in water flux and antifouling ability [7,8]. Furthermore, the presence of NZ having -OH groups on the membrane's surface influenced its hydrophilic property. The -OH groups on the surface tend to form intermolecular attraction between the water molecules, primarily hydrogen bonding.

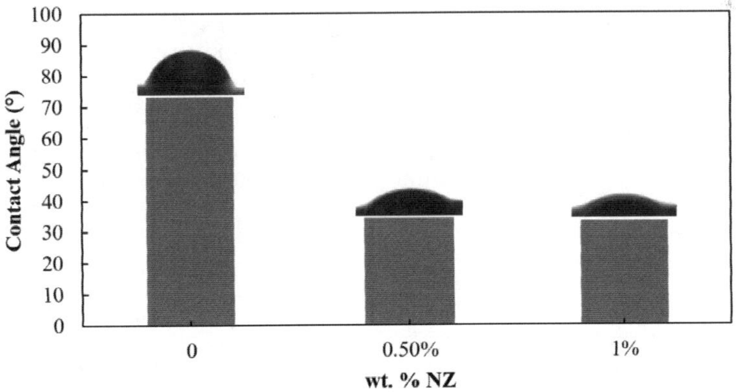

Figure 3. Contact angle measurement of the membrane blends.

3.4. Statistical analysis of the contact angle measurements

To confirm if the decrease in the contact angle measurements were statistically significant, ANOVA was conducted. The null hypothesis considered was no significant difference between the contact angle measurements obtained. Based on the result of the statistical analysis with 95% confidence level, the *p*-value was found to be less than 0.05. Thus, the decrease in the contact angle measurement due to NZ addition statistically significant. In order to determine where the significance between the measurements came from, a post hoc test was done. Based on the statistical computation, there was a significant difference between pure PCL and each of the blend. This further mean that aside from the result of the contact angle measurement, the effect of the incorporation of NZ on PCL was statistically significant. Introducing NZ on PCL matrix really caused a significant effect in terms of the membranes' wettability.

4. Conclusion

The results of the study justified the objectives of the experiment. Different PCL/NZ membrane blends with varying NZ concentration were fabricated by hand-casting method through phase inversion technique. It was found out from the SEM results that increasing the concentration of NZ on the membrane facilitated the increase in the formation of pores in the membrane. The results from the UTM showed that the presence of NZ particles on PCL polymeric matrix can improve its tensile strength by means of forming interfacial bonds. With regards to the data from the goniometer, it was found that adding NZ on PCL significantly increased the hydrophilicity of the membrane due to the hydrophilic nature of NZ. This was further supported by the statistical analysis that was conducted.

5. References

[1] Shrivastava A 2018 Environmental Aspects of Plastics *Introd. to Plast. Eng.* 207–32

[2] Mclauchlin A R and Thomas N L 2012 Biodegradable polymer nanocomposites *Adv. Polym. Nanocomposites Types Appl.* 398–430

[3] Luyt A S and Malik S S 2019 *Can Biodegradable Plastics Solve Plastic Solid Waste Accumulation?* (Elsevier Inc.)

[4] Aquino R R, Tolentino M S, Elacion R M P D, Ladrillono R, Laurenciana T R C and Basilia B A 2018 Adsorptive removal of lead (Pb2+) ion from water using cellulose acetate/polycaprolactone reinforced nanostructured membrane *IOP Conf. Ser. Earth Environ. Sci.***191** 2–10

[5] Saleh T A, Parthasarathy P and Irfan M 2019 Advanced functional polymer nanocomposites and their use in water ultra-purification *Trends Environ. Anal. Chem.***24**

[6] Irandoost M, Pezeshki-Modaress M and Javanbakht V 2019 Removal of lead from aqueous solution with nanofibrous nanocomposite of polycaprolactone adsorbent modified by nanoclay and nanozeolite *J. Water Process Eng.***32** 100981

[7] M S Tolentino, R R Aquino, M R C Tuazon, B A Basilia, M J Llana J and Cosico 2019 Adsorptive removal of Ni 2 + ions in wastewater using electrospun cellulose acetate / iron-modified nanozeolite nanostructured membrane Adsorptive removal of Ni 2 + ions in wastewater using electrospun cellulose acetate / iron-modified nanozeolite nanostr *IOP Conf. Ser. Earth Environ. Sci.***344**

[8] Amiri F, Moghadassi A R, Bagheripour E and Parvizian F 2017 Fabrication and characterization of PES based nanofiltration membrane modified by zeolite nanoparticles for water desalination *J. Membr. Sci. Res.***3** 50–6

IOP Conf. Series: Materials Science and Engineering **894** (2020) 012005 doi:10.1088/1757-899X/894/1/012005

Effect of Lithium Chloride on the Fabrication of Polyvinylidene Fluoride Nanocomposite Membrane for the Separation of Oil-Water Emulsions

Mary Rose C Tuazon[1], Ruth R Aquino[1, 2], Micah Belle Marie Yap Ang[3] and Marvin S Tolentino[1,4]

[1]School of Chemical, Biological, and Materials Engineering and Sciences, Mapua University, 658 Muralla St., Intramuros, Manila 1002, Philippines
[2]Center for Nanomaterials and Membrane Research, Mapua University, 658 Muralla St., Intramuros, Manila 1002, Philippines
[3]R&D Center for Membrane Technology and Department of Chemical Engineering, Chung Yuan University, Taoyuan 32023, Taiwan
[4]Industrial Technology Development Institute, Department of Science and Technology, Bicutan, Taguig City, Metro Manila 131, Philippines

mrctuazon@mymail.mapua.edu.ph, rraquino@mapua.edu.ph

Abstract. Polyvinylidene fluoride membranes have been broadly applied to scientific research and industrial processes due to its excellent property and applications mainly in water treatment. In this study, the effects of lithium chloride (LiCl) on the porosity, morphology, hydrophilicity and permeability of PVDF-halloysites (PVDF-HNT) nanofiber membranes were investigated. It was observed that the incorporation of LiCl improved the electrospinnability of polymersolution and modifes the morphology of fibers from beaded fiber to a uniform fiber structure. The fabricated membrane was used for the separation of oil-water emulsions via gravitational dead end system, in which the 2% concentration of LiCl as additive, bear out the most amount of volume of water collected. As a result, the fabricated membrane offers a new material for practical applications in oil-water separations.

1. Introduction

Oil spill is a type of environmental pollution. It is discharged by accident or purposely that hovers on the surface of water bodies as a distinct mass and is brought off by currents, tides, or winds. In the Philippines, various incidents of oil spill were reported due to discharge of oil by mistake or on purpose. Oil spill can be partly controlled by combustion, chemical dispersion, and adsorption. A lot of procedures have been developed to eradicate oil spillfrom polluted water. Oil-water separation is a global challenge due to increasing oily wastewater, and repeated oil spill incidents [1]. One of the solutions for dealing this concern is the reuse of water that involves the adoption of developed technologies, like membrane technology. Membrane technology has been lately launched as an effective method to separate oil-water emulsion due to its capacity to efficiently take away the oil droplets in contrast to the existing conventional method[2,3,4,5,6]. Addition of salt additives (e.g. lithium chloride) to prepare porous membranes has been an effective practice to improve permeability and selectivity.The objective of this study is to fabricate an electrospun membrane for separation of oil-water emulsions. The target is to produce a PVDF nanocomposite membrane that will act as a substrate via electrospinning technique, characterize the fabricated membrane according to surface morphology, chemical structure, mechanical properties, and hydrophilic/hydrophobic property, determine the effects of incorporating the additives lithium chloride (LiCl) and halloysite nanotubes (HNT) to the polymer solution, and lastly evaluate the membrane separation performance of oil-water emulsions using gravitational set-up.

2. Materials and methods

The materials used in this study are Polyvinylidene fluoride (PVDF, Kynar 2800), commercial halloysitenanoclays (HNT), N-Methyl-2-Pyrrolidone (NMP), lithium chloride (LiCl), distilled water, Triton x-100 (emulsifier) and engine oil. The materials are purchased from Sigma Aldrich and Department of Science and Technology. All reagents used in this study were employed with no further purification.The PVDF-LiCl/HNT solution was prepared with varying concentrations of PVDF and LiCl while the amounts of solvent NMP and HNT were held constant. A total of four blends were prepared for membrane fabrication. The composition of membrane by % weight are shown in Table 1.

Table 1. Composition of blend solution

Blend	PVDF	HNT	LiCl	NMP
M0	20	0	0	80
M1	17	2	1	80
M2	16	2	2	80
M3	15	2	3	80

For the fiber production, an electrospinning machine (NANON-010A) was used. The prepared blend solutions of around 8 mL were poured individually into a 10-mL syringe container for electrospinning process. Voltage of 20 kV and feed rate of 0.9 mL per hour are consideredinfiber production. Assessment of the produced electrospun membranes is carried out to analyze the surface morphology using Scanning Electron Microscope (SEM), surface chemical composition using Fourier Transfrom Infrared Spectroscopy (FTIR), mechanical properties using Universal Testing Machine (UTM), membrane pore size using Porometer (Porolux 1000), wettability using Contact Angle (Sessile drop method). The prepared membrane composition for electrospinning process influence the viscosity and ionic conductivity making each four blend solution electrospinnable.

3. Results and discussion

3.1.Membrane Surface Morphology

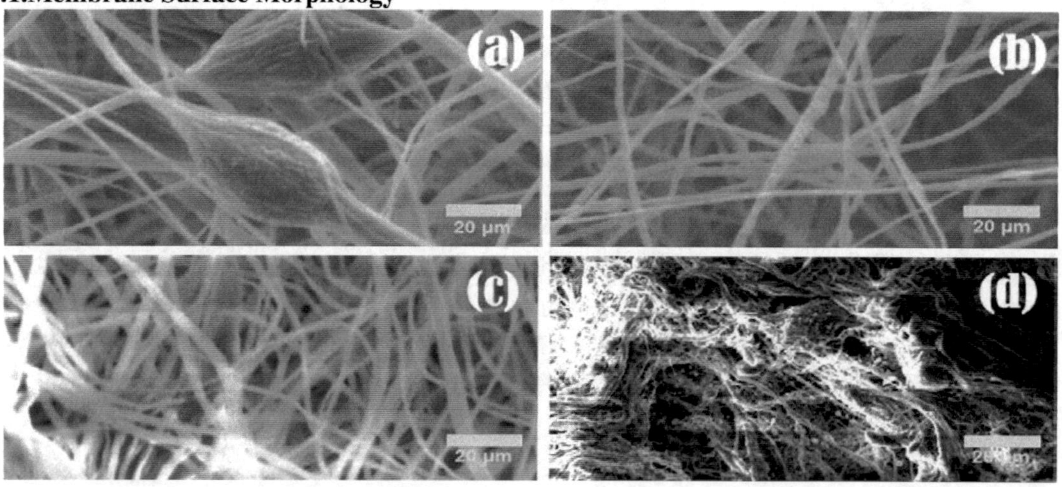

Figure 1. SEM images of (a) 0% LiCl, (b) 1% LiCl, (c) 2% LiCl and (d) 3% LiCl

Scanning Electron Microscope was used to determine the surface morphology of each of the electrospun membranes under magnification of 5000x with an accelerating voltage of 10 kV. Addition of salt in the solution was found to bring a remarkable change to the membrane morphology of the samples. The remarkable change was due to the addition of salt, lithium chloride, on the membrane, where the morphology of fibers from beaded fiber changes to a more uniform fiber structure. According to studies that addition of small amount of salt will bring about a significant decrease in fiber diameter as shown in Table 2 below which is probably due to higher conductivity from lithium chloride.

Figure 1a exhibits a minimal quantity of beaded fiber with an average fiber diameter of 508.9 ± 22 nm, but when lithium chloride was added the bead-on-string fibers gradually disappeared as shown in figure 1b with average fiber diameter of 352.2 ± 21 nm. The cause of beads formation has been reported when the concentration of solution or the molar weight of the polymer used is too low. Bead formation could be associated to lack of resistance of the polymer solutions to resist electrical force stretching due to low molecular weight or low viscosity of the polymer. In figure 1c, when the amount of lithium chloride was increased to 2%, parallel double-stranded fibers with relatively thinner diameters were achieved. Attributed to the fact that with increasing amount of lithium chloride, higher viscosity of the solution can be achieved; this led to improvement in ionic conductivity of the solutions resulting in reduced possibility of bead formation. However, upon increasing the amount of lithium chloride to 3%, the fiber structures as shown in figure 1d were obtained but with crystallization-associated morphology, it is evident with the presence of lithium chloride in the membrane but with high salt content. This is because high salt content are in meta-stable states that are highly supersaturated with respect to PVDF crystallization.

Table 2. Fiber diameter of the electrospun membranes

Membrane	Diameter (nm)
M0	508.9 ± 22
M1	432.2 ± 22
M2	352.2 ± 21
M3	308.2 ± 16

3.2.Chemical Structure

Fourier Transform Infrared Spectroscopy was used to determine the surface chemical composition. An FTIR result for PVDF nanocomposite membranes with varying concentrations of lithium chloride are illustrated in Figure 2. Decreasing intensities of spectra were observed at 1000 cm^{-1} to 1500 cm^{-1}. Distinct but very low intensities of halloysite spectra were observed at 1649 cm^{-1}that corresponds to water interlayer's OH bending.

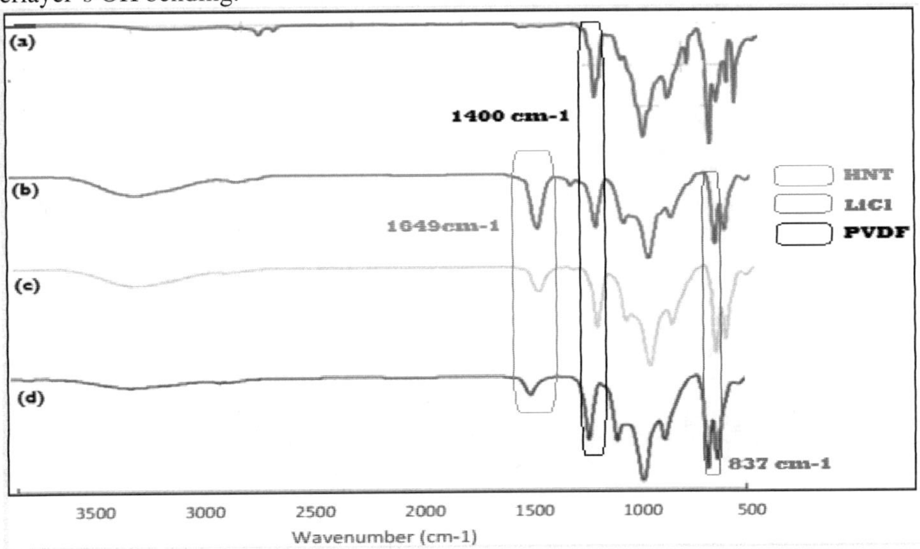

Figure 2.FTIR Spectrum for PVDF:HNT:LiCl in different ratio (a) Pure PVDF (b) 17:1:1 (c) 16:2:2 (d) 15:2:3

Absorption peaks at 872 cm^{-1} and 841 cm^{-1} from PVDF spectra are retained in the blends containing 1%, 2% and 3% LiCl. C-C-C asymmetrical stretching vibration and peak at 1400 cm^{-1} that corresponds to the C-F stretching vibration of PVDF appeared in lower intensity. From Figure 2 (b)-(d), sharp and strong peak located at 837 cm^{-1} correspond to the existence of LiCl in membranes while PVDF membrane with no additives illustrate coarse and weak peak at the same wavelength.

3.3. Mechanical Property

Since PVDF membranes wear off easily when subjected to abrasive materials, mechanical stability should be taken into account. Young's modulus and tensile strength were mainly used to evaluate the mechanical property of the produced membrane as shown in Table 3.

Table 3. Young's modulus and tensile strength of membrane samples

Sample	Tensile Strength (MPa)	Young's modulus (MPa)
M0	2.4478	2.3650
M1	3.8876	3.4234
M2	3.1609	2.1760
M3	3.0176	2.0965

It is clear that PVDF with halloysitenanoclays have higher tensile strength and Young's modulus than pure PVDF which suggest that incorporation of halloysitenanoclays gives extra stiffness to the polymer matrix. It is possible that the intrinsic modulus of halloysitenanoclays are present and are relatively easy to disperse in the polymer matrix. But, keeping the amount of halloysite constant while the amount of LiCl increases, the membrane leads to a lower mechanical stability. The possible reason is that LiCl concentration is higher than halloysite concentration. The porous structure created by lithium chloride can be the basis of the significant decrease in Young's modulus and tensile strength. There must be an optimum amount in HNT/LiCl ratio to develop strength, but excessive amount will decrease the strength.

3.4.Contact Angle

The hydrophilicity of the membranes was enhanced significantly by adding different amount of LiCl. As a result, as the amount of LiCl increases, membrane hydrophilicity also increases.As shown in Figure 3, the increase of membrane hydrophilicity was attributed to the nature of lithium chloride being highly polar. Because cation like lithium and anion like chloride can offer good membrane hydrophilicity via ion-dipole interaction, where these lyotropic salts can produce complexes with the carbonyl group in polar aprotic solvent such as N-methyl-2-pyrrolidone (NMP).

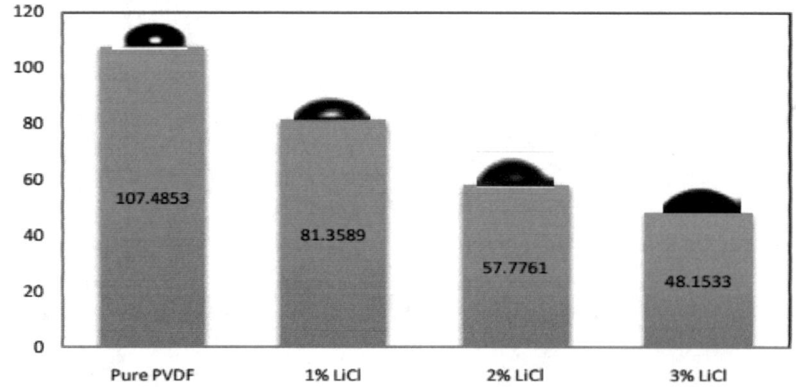

Figure 3. WCA of membranes with varying concentration of LiCl

3.5. Oil-Water Separation Performance

The four electrospun membranes underwent oil-water separation test for 24 hours. It is obvious that incorporation of LiCl showed a significant improvement on the volume of water collected. For pure PVDF membrane,nowater was collected after 24 hours this was due to the hydrophobic nature of the membrane. In the membrane with 1% LiCl, 4.3 mL of water was collected after 24 hours with no trace of oil. Membrane with 2% LiCl showed improvement in the volume of water collected by about 6.1 mL with no trace of oil after 24 hours. However, membrane with 3% LiCl after 24 hours exposure to oil-water separation test the membrane was tattered. The possible reason might be due to high content of LiCl and low tensile strength where it absorbs plenty of water which caused the breakage of the membrane. Given the results on Table 4, membrane with 2% LiCl gave the best result for membrane oil-water separation performance.

Table 4. Volume of water collected after 24 hours

Membrane	Volume of water collected (mL)
M0	0
M1	4.3
M2	6.1
M3	0

4. Conclusion

A novel PVDF oil-water separation nanocomposite membrane was effectively fabricated through the addition of lithium chloride via electrospinning technique. From the results gathered in this study, it can be concluded that lithium chloride as additive plays a significant part in improving the membrane performance of nanomaterials which changes the fiber morphology from beaded fiber to a more uniform fiber structure. Successful incorporation of lithium chloride on the membrane surface, confirmed by FTIR which showed significant peaks, can effectively solve poor dispersion in the membrane. These improvements are attributed to a decrease in roughness, which can prevent oil droplets from adsorbing or entering in the membrane surface and pores but increases water permeation rate into the membrane. Li+ and Cl- interacts strongly via ion-dipole interactions which targets the PVDF making it more porous, fibrous and hydrophilic. Therefore, the modification converts the membrane from highly hydrophobic to hydrophilic under oil when wet with water. This nano-enabled membrane study will definitely help the treatment of oily wastewater as an effective modification technique and key element for sustainable development in the field of membrane technology.

5.References

[1] J. W. Pattersonn, in Industrialized Wastewater Treatment.Butterworths, Stoneham, 2 M. MA, 2nd edition, 1985.

[2] A. Ezzati, A. Fakhru, E. Gorouhi, J. Hermia, T. Mohammadi, Ultrafiltration of water –olive oil emulsions using microfiltration, Desalination 185 (2005, 2014), 113-120, 371–382.

[3] J. Kong, L.Y. Ng, K. Li, Removal of oil from oil-water mixtures using PVDF ultrafiltration membranes, Separation &Purifation. Technology. 16 (1999) 15-93, 83–93.

[4] M. Cheryan, N. Rajjagopalan, Membrane filtration & processing of oily waste streams. Wastewater treatment and waste shrinking, J. Membrane.Science. 151 (1998) 13–28.

[5] S.R.H. Abadi, A. Reyhani, M. Hemati, F. Rekabdar, T. Mohammad. Low cost ceramic membrane performance in ultrafiltration/microfiltration of oily seawater, Desalination 265 (2011) 222–228.

[6] Singh, M.K. Puurkait. Cross-flow ultrafiltration/microfiltration of industrialized wastewater: experimental and theoretical consideration, Separation. Science.Technoly. 46 (2011) 1213–1223.

IOP Conf. Series: Materials Science and Engineering **894** (2020) 012006 doi:10.1088/1757-899X/894/1/012006

Nanoporous Carbon from Durian Peel via Hydrothermal-Carbonization and their Application in Ripening Delay of Durian

Thanat Sitthisantikul[1], Pee Poolsili[1], Jindabha Devakula[2], Anuchit Jaruwanawat[1] and Apiluck Eiad-ua[1*]

[1] College of Nanotechnology, King Mongkut's Institute of Technology Ladkrabang, Bangkok, Thailand
[2] Mr.Fruity Company Limited, Khlong nung, Khlong luang, Pathum thani, Thailand

[*]Corresponding author's E-mail: apiluck.ei@kmitl.ac.th

Abstract: This research was aimed to examine conditions on synthesis of nanoporous carbon materials from durian peel via the process of hydrothermal carbonization (HTC) as an application for delaying durian ripening process. The experiment was conducted by using durian peel as materials for producing nanoporous carbon via the process of hydrothermal at 160-200 ºC for 8-24 hours. It also included the process of carbonization at 500-900 ºC for 2 hours under nitrogen atmosphere for developing pore structure and removing contaminants to obtain the nanoporous carbon. The properties of nanoporous carbon were characterized by scanning electron microscope (SEM), Fourier transform infrared spectroscopy (FT-IR) and Raman spectrometer analysis. The results indicated that the process of hydrothermal at 200 ºC for 12 hours and carbonization at 900 ºC for 2 hours was suitable for delaying ripe durian. This is become of the carbon content, porous structure and amorphous structure increased with HTC temperature.

1. Introduction

Durian (*Durio zibethinus*) is one of the tropical fruits in Southeast Asia such as Indonesia, Malaysia and Thailand. This fruit has hard outer shell and distinctive smell [1]. Demands of durian consumption are significantly increasing. However, fast ripening process of durian during transportation becomes one of main problems that effect on the quality product when reaching the consumer. In addition, disposed of peel from durian consumption are causing of environmental pollution.

Synthesis of nanoporous carbon from durian peel, which contain many feature such as high surface area high porous volume [2], and resist to chemicals and heat in anaerobic condition [3-5]. It is one of effectively solution for reduce environmental pollution also obtain economic value from zero-cost waste and application as an ethylene adsorbent [6-7] for delaying durian ripening process. One of methods for effectively synthesize carbonaceous materials is hydrothermal carbonization (HTC) which is process of reaction that change biomass to carbon material by water molecule, heat and pressure. HTC process was explained with 4 reaction by the principle, hydrolysis, dehydration, polymerization and carbonization reaction, to destroy the bond such as lignin, hemicellulose and cellulose, which are the main component of biomass. Nanoporous carbon material prepared via HTC can achieve a higher yield of hydrothermal char, higher level of carbon and lower ash content [8-10].

This study was aimed to examine the synthesis of nanoporous carbon from durian peel via hydrothermal carbonization, including different hydrothermal and carbonization condition. The nanoporous carbon obtained was characterized in term of the porous structure, change of chemical composition and amorphous structure.

2. Material and method

2.1 Material
Durian peel powder was obtained from Mr.Fruity company limited, Khlong nung, Khlong luang, Pathum thani. Nitrogen with a purity of 99.95% was used as inert gas in the carbonization process.

2.2 Preparation of Nanoporous carbon.

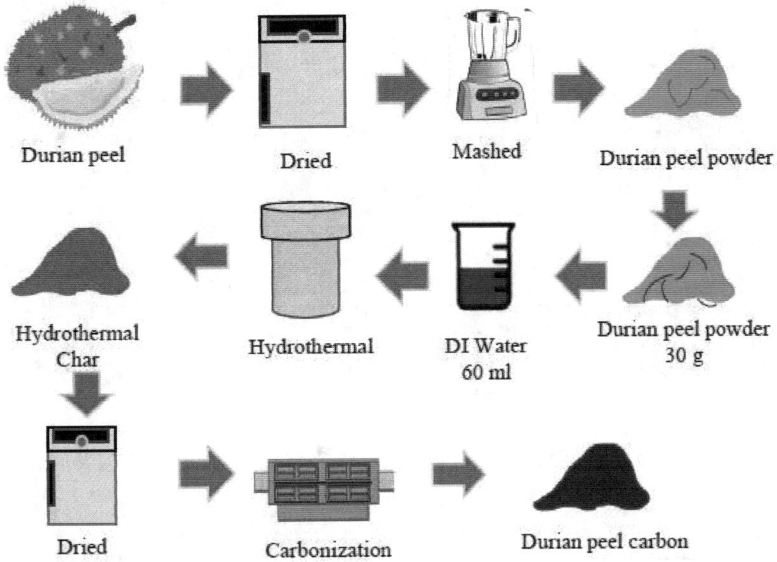

Figure 1. Schematic Diagram of Nanoporous carbon from durian peel via Hydrothermal-carbonization process.

The schematic diagram of nanoporous carbon from durian peel via HTC in Figure 1. Durian powder was dried and mashed from durian peel, 30 g of Durian peel powder was mixed with 60 ml of DI water and prepared by process of hydrothermal at 160-200 °C for 8-24 hours, afterward hydrothermal char was carbonized in furnace at 500-900 °C for 2 hours under nitrogen atmospheric to obtain nanoporous carbon.

2.3 Nanoporous carbon characterization
Nanoporous carbon from durian peel via HTC were characterized for morphology, Chemical composition and amorphous structure by scanning electron microscope (SEM), Fourier transform infrared (FT-IR) and Raman spectrometer, respectively.

3. Result and discussion

3.1 Morphology of Carbon and Hydrothermal char
Figure a-e2 shows SEM images of durian peel through HTC process at different reaction temperature (500-900 oC) under nitrogen atmospheric for 2 hours and Figure f2 show SEM image of durian peel through hydrothermal process at 200 oC for 12 hours. The SEM analysis show that increasing reaction temperature of carbonization process was destroy chemical bond structure and developing pore structure. In figure e2, hydrothermal-carbonization at 900 oC for 2 hours under nitrogen atmospheric was the optimum condition for produces ethylene adsorbent

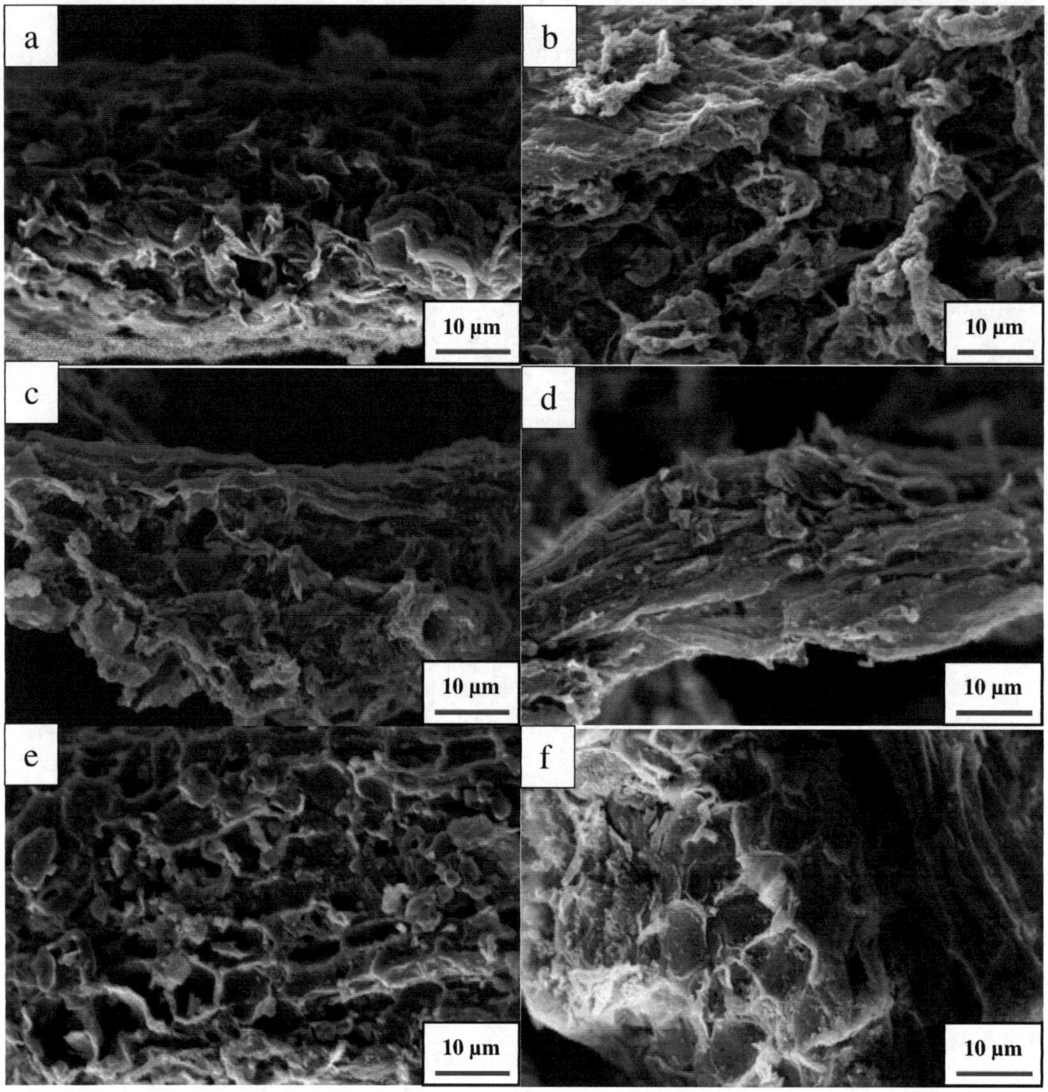

Figure 2. SEM image (500x) of durian peel through Hydrothermal process at 200 °C for 12 hours and carbonization process at different temperature of (a) 500°C (b) 600 °C (c) 700 °C (d) 800 °C (e) 900 °C for 2 hours under Nitrogen atmospheric for 2 hours and (f) Hydrothermal char at 200 °C for 12 hours.

3.2 Structure analysis and composition of porous carbon

The FT-IR spectra analysis show that in Figure a3 the peak at 3343 cm-1 was assigned to hydroxyl group, carbonyl group, alcohol from cellulose or phenols from lignin. The peak at 2918 cm-1 was represent the CH2 stretching (aliphatic). The peak at 1615 cm-1 was assigned to vibration of aromatic ring C-C stretching, which found in lignin. The peak around 1152 and 1028 cm-1 are indicated to the stretching of C-O from hemicellulose ester. The FTIR spectra of hydrothermal char from durian peel at 200oC for 12 hours are displayed in Figure b3 shows the decomposition of alkane content increased, hemicellulose, cellulose and lignin decreased. The FT-IR spectra analysis of hydrothermal-carbonization process with different reaction temperature (500-900 oC) under nitrogen atmospheric in Figure c-g3 show that OH, aliphatic C-H, aromatic C-C and C-O linkage are decreased when compared with hydrothermal char. The result indicated that decomposition of hemicellulose, cellulose, lignin and OH group are increase with increasing reaction temperature of carbonization process. The process of hydrothermal at 200oC for 12 hours and carbonization process at 900oC for 2 hours under nitrogen atmospheric was the optimum condition for produce ethylene adsorbent.

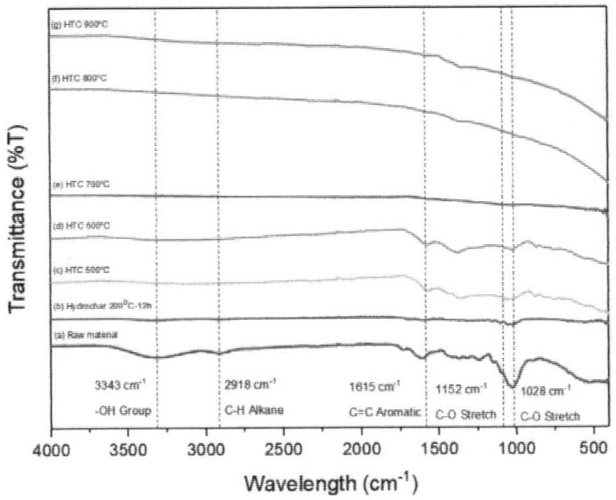

Figure 3. FT-IR spectra of durian peel through hydrothermal process at 200 °C for 12 hours and carbonization with different reaction time (a) Raw material (b) Hydrothermal char at 200 °C for 12 hours (c) HTC 500 °C (d) HTC 600 °C (e) HTC 700 °C (f) HTC 800 °C (g) HTC 900°C for 2 hours under nitrogen atmospheric.

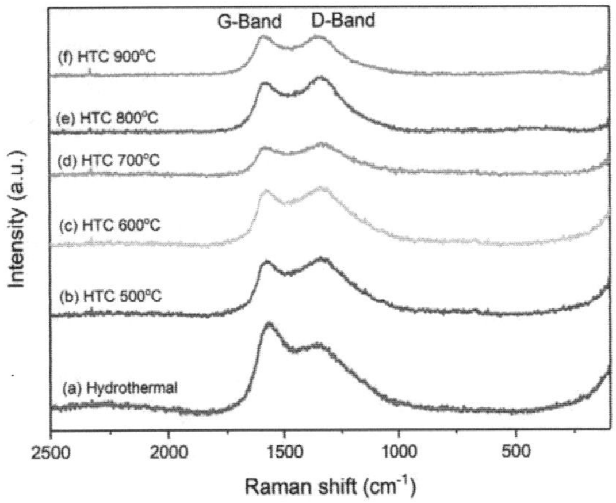

Figure 4. Raman spectra of durian peel through hydrothermal process at 200 °C for 12 hours and carbonization with different reaction temperature (a) Hydrothermal char at 200 °C for 12 hours (b) 500 °C (c) 600 °C (d) 700 °C (e) 800 °C (f) 900 °C for 2 hours under nitrogen atmospheric

Table 1. Raman spectra data of durian peel through process of hydrothermal at 200 °C for 12 hours and carbonization with different reaction temperature for 2 hours under nitrogen atmospheric

Sample	I_D	I_G	D band (cm^{-1})	G band (cm^{-1})	I_D/I_G
Hydrothermal (200 °C,12 h)	144.23	186.43	1358	1570	0.77
HTC 500 °C	122.37	114.28	1342	1577	1.07
HTC 600 °C	123.62	115.41	1339	1577	1.07
HTC 700 °C	124.07	112.51	1338	1576	1.10
HTC 800 °C	115.14	102.51	1334	1577	1.12
HTC 900 °C	66.83	47.53	1347	1582	1.40

From Raman spectra analysis with different reaction temperature of hydrothermal-carbonization process under nitrogen atmospheric in figure 4 and table 1 indicated that The ratio of I_D and I_D/I_G values for hydrothermal-carbonization for 500-900 °C were calculated to be 1.07,1.07,1.10,1.12,1.40 respectively. The result indicated that The high ratio of I_D and I_G are more carbon defect. So that, The process of hydrothermal at 200 °C for 12 hours and carbonization at 900 °C for 2 hours with the D-Band (1347 cm^{-1}) and G-Band (1582 cm^{-1}) has more carbon defect than other condition. The destroy porous structure was increasing with reaction temperature of hydrothermal-carbonization process.

4. Conclusion

The process of hydrothermal at 200 °C for 12 hours and carbonization at 900 °C for 2 hours was suitable for producing ethylene adsorbent for delaying ripe durian. This is become of the carbon content, porous structure and amorphous structure increased with reaction temperature of HTC and carbonization process

Acknowledgement
The author would like to thank The College of Nanotechnology, King Mongkut's Institute of Technology Ladkrabang and Mr.Fruity company limited for support this research.

IOP Conf. Series: Materials Science and Engineering **894** (2020) 012006 doi:10.1088/1757-899X/894/1/012006

5. Reference

[1] Chandra, T. C., M. M. Mirna, J. Sunarso, Y. Sudaryanto and S. Ismadji (2009). "Activated carbon from durian shell: Preparation and characterization." *Journal of the Taiwan Institute of Chemical Engineers* **40** (4): 457-462.

[2] Sabzoi, N., S. Shrestha, B. Si Ali and M. Siddiqui (2015). "A critical analysis on palm shell from oil palm industry as a feedstock for solid char production." *Reviews in Chemical Engineering* **32** (5).

[3] Jurewicz, K., C. Vix-Guterl, E. Frackowiak, S. Saadallah, A. Reda, J. Parmentier, J. Patarin and F. Beguin (2004). "Capacitance properties of ordered porous carbon materials prepared by a templating procedure." *JOURNAL OF PHYSICS AND CHEMISTRY OF SOLIDS* **65**: 287-293.

[4] Salunkhe, R., Y. Kamachi, N. Torad, S. M. Hwang, Z. Sun, S. Dou, J. H. Kim and Y. Yamauchi (2014). "Fabrication of symmetric supercapacitors based on MOF-derived nanoporous carbons." *Journal of Materials Chemistry A* **2** (46): 19848-19854.

[5] Yazdankhah, A., S. E. Moradi, S. Amirmahmoodi, M. Abbasian and S. Esmaeily (2010). "Enhanced sorption of cadmium ion on highly ordered nanoporous carbon by using different surfactant modification." *Microporous and Mesoporous Materials* **133**(1-3): 45-53.

[6] Haslina, E., M. Mustfha, S. Sufian and k. z. ku shaari (2013). "Methylene Blue Dye Adsorption to Durian Shell Activated Carbon." *Key Engineering Materials* **594-595**: 350-355.

[7] Yee Jun, T., P. Latif, A. Abdullah, A. Shamala-Devi and Y. H. Taufiq-Yap (2010). "Performances of toluene removal by activated carbon derived from durian shell." *Bioresource technology* **102**: 724-728.

[8] Iryani, D., S. Kumagai, M. Nonaka, K. Sasaki and T. Hirajima (2017). "Characterization and Production of Solid Biofuel from Sugarcane Bagasse by Hydrothermal Carbonization." Waste and Biomass Valorization **8**: 1941-1951.

[9] Sluiter, A., B. Hames, R. Ruiz, C. Scarlata, J. Sluiter, D. Templeton and D. Crocker (2008). "Determination of structural carbohydrates and lignin in biomass determination of structural carbohydrates and lignin in biomass." *NREL Lab. Anal. Proced.* **510**.

[10] Iryani, D., S. Kumagai, M. Nonaka, K. Sasaki and T. Hirajima (2016). "Hydrothermal

IOP Conf. Series: Materials Science and Engineering **894** (2020) 012007 doi:10.1088/1757-899X/894/1/012007

Nanoporous Carbon from Water Hyacinth Via Hydrothermal Carbonization

Sirayu Chanpee[1], Nattaya Suksai[1], Napat Kaewtrakulchai[1], Sutee Chutipaijit[1], Masayoshi Fuji[2], and Apiluck Eiad-ua[1*]

[1] College of Nanotechnology, King Mongkut's Institute of Technology, Bangkok, Thailand 10520

[2]Advanced Ceramic Center, Nagoya Institute of Technology, Tajimi, Gifu 507-0033, Japan

apiluck.ei@kmitl.ac.th

Abstract. Nanoporous carbon materials have been successfully synthesized from water hyacinth via hydrothermal carbonization (HTC). This research was studied the effect of hydrothermal temperature from 160 - 200 °C and reaction time for 4 - 24 h. Afterwards, carbonization was carried out at the temperature of 600 - 900 °C for 2 h in N_2 atmosphere for developing porosity and even removing contaminants of hydrothermal char to obtain the porous carbon. The physico-chemical properties of nanoporous carbon materials were comprehensively characterized through Scanning electron microscope (SEM), Fourier transforms infrared spectroscopy (FT-IR), CHN elemental analysis, X-ray diffraction (XRD) and BET analysis. The adsorption capacity and carbon content of nanoporous carbon materials from water hyacinth were increased with increased hydrothermal carbonization temperature and time. Performing HTC at 200 °C for 12 h. Is the optimum condition to synthesis of precursor materials for good adsorbent.

1. Introduction

Agricultural wastes are generated as unused materials from harvesting, processing and storing activities. Corn cob, bagasse and husk are examples of agricultural wastes called "biomass" [1]. Biomass mainly consists of three components- cellulose, hemicellulose and lignin- which very ideal for producing carbon materials [2]. Latterly, these conventional biomasses were used as direct fuel, resulted in higher prices and scarcity. Therefore, alternative biomasses from aquatic plants such as water hyacinth is being considered. Water hyacinth (*Eichhornia crassipes*) is one of invaded species found in many countries in Southeast Asia. It spreads quickly and aggressively, leads to uncontrollable environmental problems- flow blockages, altering water tint and odor [3-7]. Water hyacinth is an interesting alternative to produce nanoporous carbon and has many uses such as capacitors or adsorbents. With nanoporous carbon production methods [8, 9] there are many ways to choose from to match nature of raw materials and properties of the desired nanoporous carbon. In this research, water hyacinths were used as precursors to produce adsorbents. By using the hydrothermal carbonization process [10, 11]. In a temperature range of 160 to 200 °C which is a carbonization method. It uses water molecules, heat, and pressure to break bonds in organic polymers. Water molecules will expand the pores [12-17]. It is an effective way to achieve high yield of carbon materials using less energy to process. In this study, we studied the effect of temperature And the time it takes for HTC reaction, which can be observed from the morphological structure and surface characteristics using scanning electron microscopy (SEM), Fourier transform infrared spectroscopy (FT-IR), Raman spectroscopy, CHN elemental analyzer, BET analysis. Carbon materials then applied by organic substances adsorption such as paraquat and methylene blue.

2. Material and methods

2.1. Material
Water hyacinth were collected from natural canal in Ladkrabang, Bangkok, Thailand.

2.2. Method
Synthesize of porous carbon materials from water hyacinth starts from drying collected raw water hyacinth by sun exposure to expel moisture and grind it into raw powder. Prepare 30 g of raw powder with 90 mL of deionized water into Teflon canister and then installed in stainless steel reactor. Hydrothermal reaction was performed at different temperatures (160, 180, and 200 °C) and times (4, 8, 12, and 24 h). After finishing hydrothermal process, remove hydrochars from the reactor and then dried in over at 90 °C overnight. Dried hydrochars were carbonized in horizontal tube furnace at 500 – 900 °C for 2 h under constant nitrogen gas flow. Final products were analysed for chemical composition and morphology.

2.3. Characterization Techniques
The samples were characterized for surface morphology by Scanning Electron Microscope (SEM). Fourier Transform Infrared Spectroscopy (FT-IR) was used to determine the function group on surface of sample. Crystallinity was measure by using X-ray diffraction (XRD) and element was measure by using CHN elemental analyzer, respectively.

3. Results and discussion

3.1. Scanning Electron Microscope (SEM)
Morphology results from scanning electron microscope of hydrochars derived from HTC process are shown in Figure 1.

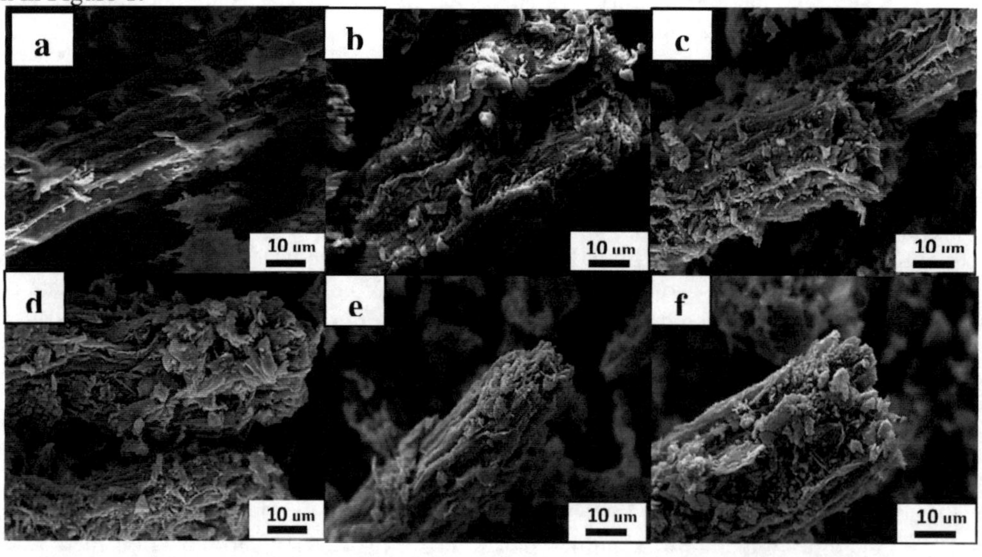

Figure 1. SEM images (500×) of Water Hyacinth hydrothermally treated at (a) 200 °C for 12 h, HTC at different conditions: (b) 500 °C, (c) 600 °C, (d) 700 °C, (e) 800 °C, and (f) 900°C.

Scanning electron micrographs of the surface morphology of the feedstock and different reaction temperatures 200 °C for 12 h are given in Figure 1(a) Carbonization at 500 – 900 °C for 2 h under N_2 Figure 1(b-f) after HTC treatment significant changes on the surface morphology can be found at different temperature. When increase temperatures provide polymer bond are more break compared to the feedstock. At temperature of 200°C showed that the hemicellulose was destroyed significantly because water molecules can diffuse through the structure dramatically. This structure implies that the formation pathway of HTC may be attributed to hemicellulose hydrolysis followed by polymerization reaction since both stage temperatures are less than 230 °C. Therefore, at temperatures below 230 °C, most of the decomposition is hemicellulose, followed by cellulose and lignin, respectively. Temperature has a significant effect on the fracture structure over reaction time. At temperature 200 °C for 12 h and carbonization at 900 °C for 2 h under N_2 is optimum condition for surface improvement of HTC which result in increased surface area and porosity.

3.2. Fourier Transform Infrared Spectroscopy (FT-IR)

The FT-IR spectra of hydrochars derived from hydrothermal process are shown in Figure 2.

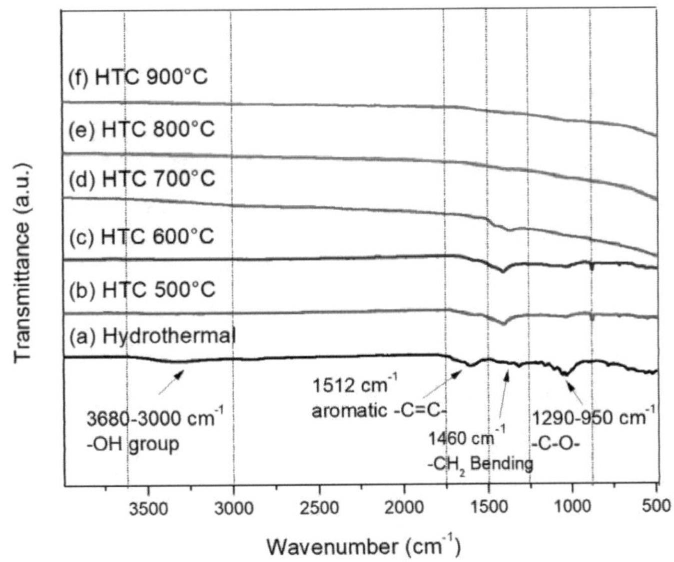

Figure 2. FT-IR spectra of (a) Water hyacinth through hydrothermal process at 200 °C for 12 h and difference co-solvent treatment, (b) 500 °C, (c) 600 °C, (d) 700 °C, (e) 800 °C, and (f) 900 °C.

A broad peak at around 3600 - 3000 cm^{-1} confirms the presence alcohols from cellulose or phenols from lignin and hydroxyl or carboxyl groups. The peak at 2925 cm^{-1} is due to CH_2 stretching vibrations of the aromatic methyl groups or aliphatic. Peaks around 1700 cm^{-1} are attributed to the stretching vibration of C=O of from cellulose and lignin [18-19]. The peak at 1600 and 1512 cm^{-1} is due to the vibration of the aromatic ring C=C stretching present in the lignin. The peak at 1460 cm^{-1} is due to CH_2 bending vibrations of the C-H deformation in lignin and carbohydrates. The absorption band observed at 1290 - 950 cm^{-1} is assigned to the stretching vibrations of C-O in aliphatic ethers and aliphatic alcohols.

3.3. CHN elemental analyzer

Elemental compositions of hydrochar and carbonization from water hyacinth at 200 °C for 4 h. and carbonization at 500 - 900 °C shown in Table 1. Carbon content in water hyacinth was about 32%, In the other head, Carbon content also increased with carbonization temperature for 45 – 48% Reductions in oxygen and hydrogen content for HTC products compared with those of water hyacinth were

IOP Conf. Series: Materials Science and Engineering **894** (2020) 012007 doi:10.1088/1757-899X/894/1/012007

observed. The reason may be that an intermolecular dehydration reaction of large molecules took place, transforming elemental hydrogen and oxygen to water.

Table 1. CHN elemental analyzer of hydrochar and carbonization samples.

	Ultimate analysis (%)			
Sample	C	H	N	O
Hydrothermal (200 °C, 12 h)	40.43	3.57	1.07	54.93
HTC 500 °C	46.04	1.80	1.33	50.83
HTC 600 °C	47.19	0.08	1.16	51.57
HTC 700 °C	46.66	-0.15	1.05	52.44
HTC 800 °C	45.49	-0.57	1.11	53.97
HTC 900 °C	48.34	-0.63	1.15	53.97

3.4. X-Ray Diffraction (XRD)

The XRD measurement was investigated to examine the phase structure of carbon materials [20], as shown in Figure 3. The broad diffraction at a 2θ angle around 19 - 26°, the broad characteristic feature indicates the major presence of amorphous phase. In addition, the peak is found at 2 theta positions of 28° and 40° was evidently noticed for all samples corresponding to the carbon [18].

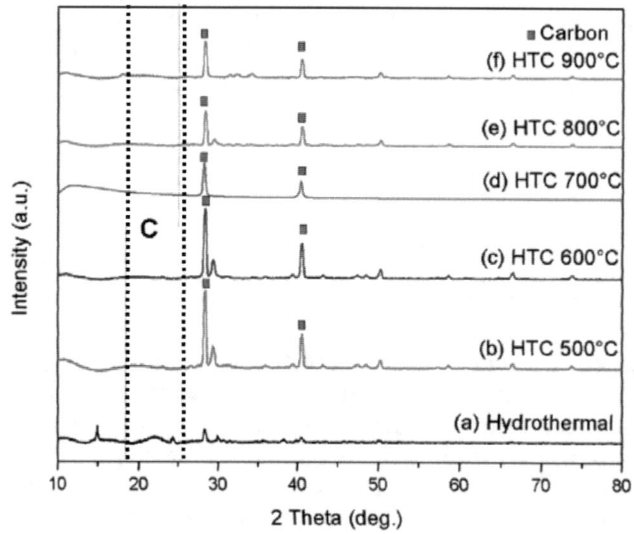

Figure 3. XRD of (a) Water hyacinth through hydrothermal process at 200 °C for 12 h and difference co-solvent treatment, (b) 500 °C, (c) 600 °C, (d) 700 °C, (e) 800 °C, and (f) 900 °C

4. Conclusions

Nanoporous carbon were successfully prepared from water hyacinth via hydrothermal carbonization at 200 °C for 12 h and carbonization at 500 - 900 °C in nitrogen gas for 2 h to develop porous structure and carbon content. It was found that carbonized hydrochar at 900 °C exhibited the highest porosity and carbon content. The temperature are mainly affected the solid yield, the chemical and structural properties of HTC. With the increasing of reaction temperature HTC yield decreased. HTC derived from decomposition of hemicellulose, residual of cellulose and formation of amorphous carbonaceous compounds during HTC. While cellulose and lignin are thermally stable components. It decomposes when hydrothermal at higher temperatures. The influence of thermal decomposition process parameters improved product characteristics and provided HTC process a promising technology may be used for various applications.

5. Acknowledgment

The authors are thankful to the College of Nanotechnology, King Mongkut's Institute of Technology Ladkrabang for supporting this work.

6. References

[1] A. Ronix, O. Pezoti, L.S. Souza, I.P.A.F. Souza, K.C. Bedin, P.S.C. Souza, T.L. Silva, S.A.R.Melo, A.L. Cazetta, and V.C. Almeida, JECE. 1830, 5(5): p. 4841-4849 (2017).
[2] A. Ausavasukhi, C. Kampoosaen, and O. Kengnok, J. Clean. Prod., 134: p. 506-514 (2016).
[3] Y. Huang, S. Li, J. Chen, X. Zhang, and Y. Chen, J. APPL SURF SCI, 293: p. 160-168 (2014).
[4] S. Nizamuddin, S. Shrestha, S. Athar, B.S. Ali, and M.A. Siddiqui, REV CHEM ENG. ISSN, 32(5) (2016).
[5] P. Regmi, J.L. Garcia Moscoso, S. Kumar, X. Cao, J. Mao, and G. Schafran, J Environ Manage, 109:p. 61-9 (2012).
[6] M. Reza, T., J. Andert, B. Wirth, D. Busch, J. Pielert, J.G. Lynam, and J. Mumme, J BIOBASED MATER BIO, 1(1) (2014).
[7] L. Wang, and F.-S. Xiao, Green Chem., 17(1): p. 24-39 (2015).
[8] M. Islam, A., M.J. Ahmed, W.A. Khanday, M. Asif, and B.H. Hameed, J ENVIRON MANAGE., 203 (Pt 1): p. 237-244 (2017).
[9] J. Xiao, J. Zhang, W. Lv, Y. Song, and Q. Zheng, CARBON, 123: p. 354-363 (2017).
[10] K. Jurewicz, C. Vix-Guterl, E. Frackowiak, S. Saadallah, M. Reda, J. Parmentier, J. Patarin, and F. Béguin, J. Phys. Chem. Solids, 65(2-3): p. 287-293 (2004).
[11] B. Liu, H. Shioyama, H. Jiang, X. Zhang, and Q. Xu, Carbon, 48(2): p. 456-463 (2010).
[12] A. Allwar, H.Z. Febriyantr, and R. Yuliantari, AJAS, 11(1): p. 20-28 (2018).
[13] N. Berge, D., K.S. Ro, J. Mao, J.R. Flora, M.A. Chappell, and S. Bae, J ENVIRON SCI, 45(13): p. 5696-703 (2011).
[14] I. Erabee, K., A. Ahsan, A.W. Zularisam, S. Idrus, N.N.N. Daud, T. Arunkumar, R. Sathyamurthy, and A.E. Al-Rawajfeh, Eng. J., 21(5): p. 1-14 (2017).
[15] Z. Liu, A. Quek, S. Kent Hoekman, and R. Balasubramanian, FUEL., 103: p. 943-949 (2013).
[16] C. Purnomo, D. Castello, and L. Fiori, APPL SCI-BASEL, 8(3): p. 331 (2018).
[17] M. Sevilla, and A.B. Fuertes, Carbon, 47(9): p. 2281-2289 (2009).
[18] J. Schwan, S. Ulrich, V. Batori, H. Ehrhardt, and S.R.P. Silva, J. Appl. Phys., 80(1): p. 440-447 (1996).
[19] L. Zhang, Q. Wang, B. Wang, G. Yang, L.A. Lucia, and J. Chen, ENERG FUEL., 29(2): p. 872-876 (2015).
[20] S. Guo, X. Dong, T. Wu, F. Shi, and C. Zhu, J ANAL APPL PYROL, 116: p. 1-9 (2015).

IOP Conf. Series: Materials Science and Engineering **894** (2020) 012008 doi:10.1088/1757-899X/894/1/012008

N-doped Porous Carbon from Palm Male Flower via Hydrothermal Carbonization

Panupong Verasarut[1], Tanatorn Liamprawat[1], Napat Kaewtrakulchai[1], Sutee Chutipaijit[1], Gasidit Panomsuwan[2], Pramote Puengjinda[3], Masayoshi Fuji[4] and Apiluck Eiad-ua[1*]

[1]College of Nanotechnology, King Mongkut's Institute of Technology, Bangkok, Thailand
[2]Department of Materials Engineering, Faculty of Engineering, Kasetsart University, Bangkok, Thailand
[3]PTT Innovation Institute (InI), Petroleum and Petrochemical Process Technology Research, Ayutthaua, Thailand
[4]Advanced Ceramic Center, Nagoya Institute of Technology, Tajimi, Gifu 507-0033, Japan

apiluck.ei@kmitl.ac.th

Abstract: N-doped porous carbon materials were produced from palm male flower using hydrothermal carbonization processes at 200 °C for 24 h followed by N-Doping and carbonization at 700°C for 2 h. N-doping was carried out by impregnation using NH_4OH at 0.5, 1.0, 1.5 M and 2 M. Products were characterized by means of chemical composition and morphology using SEM, XPS, and XRD to characterize specific properties such as physical morpholog, porosity, elemental composition on surface and crystalline structure of PMF. After applying hydrothermal carbonization processes, the results showed substantially increased porosity and surface area with suitable microstructure for N-doped electrodes applications. The highest porosity was obtained at NPC–1.5 M.

1. Introduction

Oil palm industry is one of main economic activities in Thailand. Overgrown of palm male flower become a problem which decrease productivity due to unbalanced proportion between male and female flower. Oil palm planters then require removing this male flower and become agricultural waste. Basically, planters use cut male flower as fertilizer. However, male flowers have fascinating potential to transform to carbon materials because they have organic structure as same as other biomass [1, 12]. Resulting carbon materials can be applied for various applications such as activated carbon and carbon fiber. Using wasted palm flower can help protecting the environment by reducing wasted materials and increase valve on agricultural by-products [2, 3].

Thermochemical process is the main activity in order to transform biomass such as chemical vapor deposition, hydrothermal treatment and pyrolysis or carbonization are examples of thermochemical process [4]. These process porous carbon material which shows adsorptive, mechanical and electrical properties, suitable for novelty applications. Porous carbon materials are increasing in popularity of environmental and energy utilizations [13, 14, 15]. Due to porous carbon's high porosity, it's perfectly used as electrodes. Unfortunately, electrochemical properties are limited by charge accumulation mechanics [5, 6]. Therefore, the study introduced N-doping with porous carbon derived from palm male flower. The procedures are divided to 2 steps – hydrothermal treatment and doping of nitrogen using NH_4OH solution. Another step is carbonization process to generate porous structures at high temperature and inert gas flow. Furthermore, characterization by means of morphology and chemical compositions are carried out to investigate and confirm the resulting products for further applications.

2. Material and method

2.1 Material
Palm male flower was obtained from cultivator in south of Thailand.

2.2 Preparation of N-doped Porous Carbon from palm male flower
Palm male flower (PMF) that used as feedstock for synthesis of N-doped Porous Carbon (NPC). Firstly, the feedstocks were crushed and sieved into the size of 800 μm, then, dried in an electrical oven at 90 °C for overnight. 20 g of prepared PMF was mixed with 40 g of several concentrations of ammonium solution (0.5, 1, 1.5, 2 M). Subsequently, the slurry was packed into a stainless-steel autoclave, and heated to 200 °C for 24 h to serve a nitrogen doping. Afterwards, the solid product was quenched via water and dried at 100 °C until dried. Finally, PMF-NPC were obtained from the carbonization at 700°C for 2 h. Furthermore, the prepared NPC was washed several times with deionized water for cleaning up impurities and then dried and kept for further characterization.

2.3 NPC characterization
N-doped porous carbon from palm male flower via hydrothermal carbonization were characterized for physical morphology, porosity, elemental composition, and crystalline structure by scanning electron microscope (SEM), X-ray photoelectron spectroscopy (XPS) and X-ray diffractometer (XRD), respectively.

3. Results and discussion

3.1 Physical Morphology of N-doped Porous Carbon (NPC)

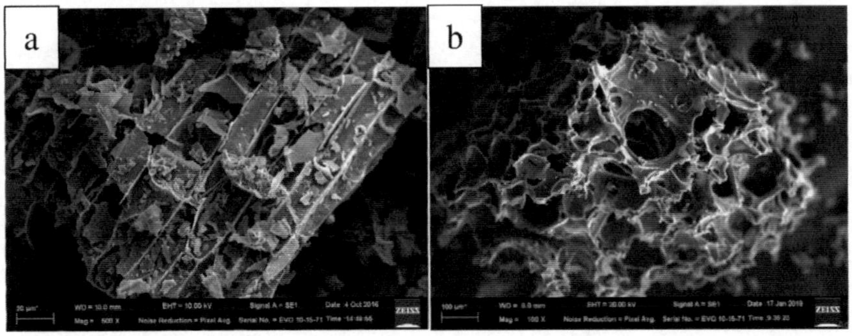

Figure 1. SEM micrographs of NPC products under 1.5 M of ammonium solution a) side view and b) top view.

Figure 1 shows the SEM micrographs of NPC obtained from hydrothermal and carbonization approaches. The surface of NPCs at the ammonia concentration of 1.5 M was best condition, as illustrated in Figure 1a and b, respectively. This is because ammonia was served as activating agent. Moreover, results indicate that the decomposition reactions occurred during the carbonization had led to the production and release of volatiles from the raw feedstock, and the activation was occurred simultaneously remaining the NPC product with high porosity.

3.2 Elemental Composition of N-doped Porous Carbon (NPC)

Table 1. Elemental composition of NPC conducted by X-ray photoelectron spectroscopy, XPS.

Conditions	Elemental compositions (%)		
	N 1s	O 1s	C 1s
Palm male flower, PMF	1.072	23.465	75.462
PMF-0.5 M	1.202	22.449	76.348
PMF-1 M	1.650	20.732	77.616
PMF-1.5 M	2.504	20.271	77.225
PMF-2 M	2.845	18.626	78.529

XPS was conducted to confirm the elemental composition on NPCs surface. As list in table 1, the carbon contents are 75.462 – 78.529 at. %, the oxygen contents are 18.626 – 23.465 at. % in PMF and PMF-NPC, respectively. The existence of oxygen in the surface of the PMF-NPC is an oxygen atom in biomass structure [1, 3, 5]. The nitrogen functional groups that usually come from the doping agents and precursors tends to increase by concentration of NH_4OH.

3.3 Crystalline structure of N-doped Porous Carbon (NPC)

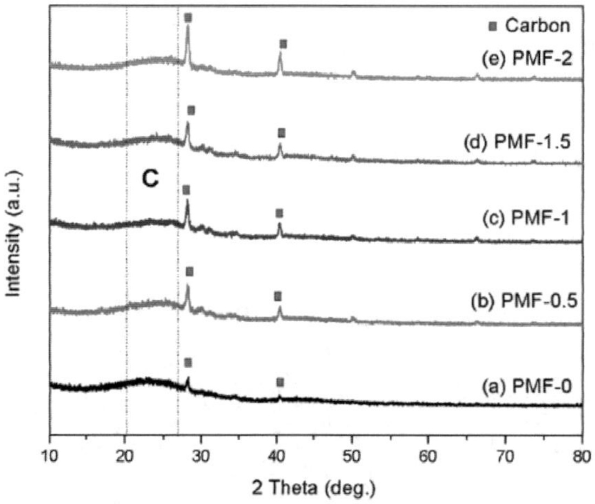

Figure 2. XRD spectra of palm male flower under conditions of several concentrations of ammonium solution a) 0, b) 0.5, c) 1, d) 1.5 and e) 2 M.

XRD measurement was investigated to examine the phase structure of NPC after activation with NH_4OH, as shown in Figure 2. The broad diffraction at a 2θ angle around 10 - 80° was evidently noticed for all samples, corresponding to the carbon. The broad characteristic feature indicates the major presence of amorphous phase. In addition, the peak is found the structure of carbon and nitrogen that may be left from activation of NH_4OH after washing.

4. Conclusion

NH_4OH solution decorated in hydrothermal treatment explores the efficient technique for synthesis of nitrogen-doped porous carbon as a novel carbon material. However, the combination step using carbonization could be implied that the pore formation and purification of pure carbon have been existed. The nitrogen percentage of NPCs are significantly enhanced with an increasing of concentration of NH_4OH. Besides, the nitrogen contained in carbon structure, their porosity and carbon content were also developed since ammonium molecules on carbon surfaces assisted an activation under a high temperature during carbonization, simultaneously. The highest porosity was obtained at NPC–1.5 M. Regarding vast physicochemical characteristics (*i.e.* huge surface area, containing N atom, high C composition), it can be implied that produced NPC could be further applied in various areas of electronic devices.

Acknowledgement
The authors would like to acknowledge to the College of Nanotechnology, King Mongkut's Institute of Technology Ladkrabang and Department of Materials Engineering, Faculty of Engineering, Kasetsart University for their assistance for facility supports.

5. Reference

[1]Ana, K.; Jelena, D.; Tatjana, T.P.; Marija, V.; Maja, P.; Vesna, M.; Zlatko, R.; Zoran, L., "Preparation of boron-doped hydrothermal carbon from glucose for carbon paste electrode" Carbon95, 2015, 42-50.

[2]Hasan, S.; Fuat, G.," High surface area mesoporous activated carbon from tomato processing solid waste by zinc chloride activation: process optimization, characterization and dyes adsorption" Journal of Cleaner Production, 113, 2016, 995-1004.

[3]Imran, J.R.; Rajalakshmi, N.; Dhathathreyan, K.S.; Ramaprabhu, S.," Nitrogen doped graphene prepared by hydrothermal and thermal solid state methods as catalyst supports for fuel cell" international journal of hydrogen energy 2015, 1-12

[4]Melike, D.O.; Uday, K.V.; Claudiu, T.L.," Preparation of industrial sisal fiber waste derived activated carbon by chemical activation and effects of carbonization parameters on surface characteristics" Industrial Crops and Products, 95, 2017 583–590..

[5]Won-Hee, L.; Jun, H.M.," Monodispersed N-Doped Carbon Nanospheres for Supercapacitor Application" ACS Appl. Mater. Interfaces, 2014, 6, 13968−13976.

[6]Latham, K.G.; Jambu, G.; Joseph, S.D.; Donne, S.W.," Nitrogen Doping of Hydrochars Produced Hydrothermal Treatment of Sucrose in H2O, H2SO4, and NaOH" ACS Sustainable Chem. Eng. 2014, 2, 755−764..

[7] Huagen Liang et al. "Highly performing free standing cathodic electrocatalysts for Li-O2 batteries: CoNiO2 nanoneedle arrays supported on N-doped carbon nanonet, Applied Catalysis" A, General 574 (2019) 114–121

[8] Qianzhi Gou et al. "Hierarchical structured porous N-doped carbon coating MnO microspheres with enhanced electrochemical performances as anode materials for lithium-ion batteries", Electrochimica Acta 296 (2019) 730e737

[9] Fangyan Liu et al. "Highly microporous carbon with nitrogen-doping derived from natural biowaste for high-performance flexible solid-state supercapacitor", Journal of Colloid and Interface Science 548 (2019) 322–332

[10] Xing Tong et al. "Tailoring the physicochemical properties of chitosan-derived N-doped carbon by controlling hydrothermal carbonization time for highperformance supercapacitor application", Carbohydrate Polymers 207 (2019) 764–774

[11] Qiong Wu et al. "N-doped porous carbon from different nitrogen sources for highperformance supercapacitors and CO2 adsorption", Journal of Alloys and Compounds 786 (2019) 826e838

[12] Gege Huang et al. "High-performance hierarchical N-doped porous carbons from hydrothermally carbonized bamboo shoot shells for symmetric supercapacitors", Journal of the Taiwan Institute of Chemical Engineers 96 (2019) 672–680

[13] Litao Yan et al. "Biomass derived porous nitrogen doped carbon for electrochemical devices", Green Energy & Environment 2 (2017) 84-99

[14] Guoyu Zhong et al. "Calcium Chloride Activation of Mung Bean: A Low-Cost, Green Route to N-Doped Porous Carbon for Supercapacitors", ChemistrySelect 2019, 4, 3432 –3439

[15] Yan-Qiu Lei et al. "Morphology evolution, formation mechanism and adsorption properties of hydrochars prepared by hydrothermal carbonization of corn stalk", J. Name., 2013, 00, 1-3

IOP Conf. Series: Materials Science and Engineering **894** (2020) 012009 doi:10.1088/1757-899X/894/1/012009

Synthesis and Surface Characterization of a Novel Plasma-polymerized (α-Pinene)-based Thin Film

Karl Vincent Alvarez*, Bryle Eusebio, Jonard Jairo Reyes, and Jason Pechardo

Department of Mining, Metallurgical, and Materials Engineering, University of the Philippines Diliman, Quezon City, Philippines

knalvarez@up.edu.ph

Abstract. α-Pinene is a bicyclic compound that is commonly sequestered from coniferous plants like pine trees. Several properties of pinene have been investigated, such as its isomerization capabilities and biological activity. However, the viability of pure pinene as an organic thin film has not yet been explored. In this work, we report the synthesis of an α-pinene-based organic thin film deposited on Al and glass substrates via plasma polymerization and its subsequent surface characterization. It was determined that the film generally became more hydrophobic as the α-pinene loading and the deposition time increased. More importantly, it was shown that specifying a pinene-Ar ratio of 50:50 yielded the highest contact angles at 94.22° and 110.38° for deposition times of 6.0 and 9.0 minutes, respectively. These results signify that a pinene-based organic thin film has been produced through the plasma polymerization synthesis route.

1. Introduction

The thin film industry has progressed considerably to adapt to the current needs and trends of the market. For major thin film industries, such as microelectronics, the thin film is typically sourced from metals and metal oxides. Thin films have also been reported to be suitable for pharmaceutical and biomedical applications, such as thin-film drug delivery and bioactive coatings for prosthetics. [1,2] While some inert and known biocompatible metals (e.g. Ti) have been routinely used for such applications, there is still a need to provide an alternative thin film that offers environmental sustainability, biocompatibility, and cost-efficiency.

To this end, the research for organic thin films have gained traction in the recent years. Due to its composition, organic thin films can naturally degrade or decompose, thereby promoting environmental sustainability. [3]

It is with this motivation that Gerchman, et al. developed a thin film based on the organic compound limonene. [4] Limonene is a member of the terpene class of compounds—organic substances that are cycloaliphatic or aromatic in nature. [4,5] The cost of its production is cheap, as it is often sourced from the by-products of the citrus production industry. i.e. from waste peels of fruits such as oranges and lemons.

Another potential compound that can be formed into a thin film is α-pinene. α-Pinene ($C_{10}H_{16}$), as shown in Figure 1, is a monoterpene organic compound typically extracted from the resin of pine trees and other coniferous flora. [6] Literature indicates that pinene has been known to exhibit antimicrobial and anti-inflammatory properties, with the (+)-α-pinene enantiomer showing promise as a human anti-osteoarthritic drug. [7]

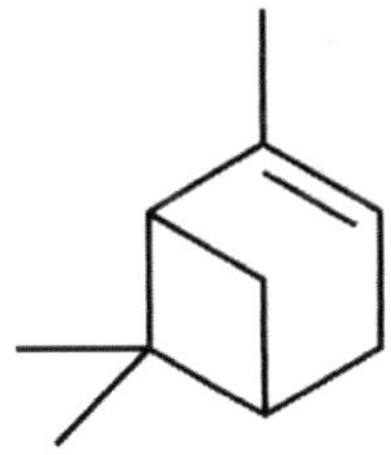

Figure 1. Structure of α-Pinene, the more abundant form of pinene in nature

Conventional polymerization routes prove to be difficult for compounds such as limonene and pinene because of the low conversion rates and its inefficiency arising from degradative chain transfer reactions. [8] Instead, the use of plasma polymerization affords the production of thin films from any organic precursor. Utilizing a stable plasma between two electrodes produces a myriad of reactive species like electrons, ions and radicals that bombard the organic molecule. [4,8] These highly energetic interactions initiate bond cleavage in the organic precursor, resulting in the fragmentation, rearrangement, and eventual crosslinking of the latter onto the substrate. [4,9]

Apart from its versatility, the plasma polymerization route is advantageous because of its ease of use. Several studies, prior to the work of Gerchman, et al., have reported the usage of plasma polymerization in creating organic thin films out of terpenoid substances. The work of Bazaka, et al., for instance, synthesized terpinen-4-ol using radio frequency (RF) plasma polymerization technique. [10] Pegalajar-Jurado, et al. and Chan, et al. have likewise used RF plasma polymerization to produce thin films based on 1,8-cineole and carvone, respectively, and explore their antimicrobial and biocompatibility properties. [11,12] While pinene belongs to the same family of compounds as the others previously mentioned, there have been no prior studies dealing with its synthesis and utilization as a thin film.

With these in mind, the present work is primarily geared towards synthesizing an (α-pinene)-based organic thin film via plasma polymerization. Subsequently, the study aims to characterize the surface morphology of the thin film formed and explore the effects of the deposition parameters, specifically the deposition time and the (α-pinene)-Ar gas ratio, on the said properties. However, due to constraints in time and resources, additional surface characterizations were not employed.

2. Methodology

2.1. Sample Preparation

Eighteen samples of 1 x 1-inch Al slabs were grinded with 1500 and 2000 grit size sandpaper. Then, the metal sheets were subjected to a series of cleaning in different solvents (deionized water, ethanol, acetone) via ultrasonication for 30 minutes. For characterization purposes, the same washing process was done to three 1 x 1-mm glass slides.

2.2. Plasma Polymerization

Deposition of pinene thin film was done via plasma polymerization in a customized RF plasma system in Plasma-Material Interactions Laboratory (PMIL). For the plasma polymerization of pinene, a base pressure of 10-15 Pa was first attained before introducing Ar and α-pinene (British Drug Houses Ltd.) gases. RF power was then set at 50 W to initiate plasma formation. Two levels for the deposition time were used (6.0 and 9.0 minutes). Meanwhile, three levels for the gas ratios were used: 25-75, 50-50, and 75-25 percent pinene-to-argon. The gas ratios were measured via partial pressure, i.e. 25-75% pinene-to-argon implies the use of 25 Pa of pinene to 75 Pa of Ar for the plasma polymerization process.

2.3. Characterization

Sessile drop test method was conducted to each treated Al samples by dropping 5.0 μL of deionized water onto the metal surface using a micropipette. The droplets' images were captured using a high magnification camera (DinoCapture 2.0) and the contact angle was processed using ImageJ.

Surface analysis, including surface morphology and homogeneity, was done via FESEM (InTouchScope, UP Baguio) in SEI mode with an accelerating voltage of 10 keV and a working distance of 19 mm. Analysis of the chemical composition of the plasma polymerized α-pinene was done via FTIR-ATR ranging from 4000 cm^{-1} and 500 cm^{-1}.

3. Results and Discussion

Shown in Figure 2 below is the trend for the average H_2O contact angles obtained for the plasma-polymerized pinene thin films. As seen in the figure, there was a consistent increase in hydrophobicity of the treated Al surface as the deposition time was increased from 6.0 to 9.0 mins. The said trend occurred for all the samples with constant pinene-Ar gas ratio but varied time of exposure. The phenomenon observed implies that heightened exposure time deposited more α-pinene layers as compared to that with minimum time of deposition since prolonged deposition amplifies the probability that the ionized solvent would interact with the metal surface.

Using the same figure, it can be noticed that the trend observed for varying gas ratios deviated from ideality. In line with Le Chatelier's Principle, it was expected that additional amounts of the organic reactant would eventually result to a superior hydrophobicity of the surface. [13] However, the results showed that the best gas ratio was at 50% pinene and 50% Ar. It can be inferred that the lowest setting was governed by Ar sputtering. On the other hand, the highest pinene-Ar ratio resulted to insufficient amount of Ar gas that will ionize the bulky organic precursor and carry it to the substrates. Thus, the lack of ionizing gas possibly resulted to the inefficient deposition of α-pinene.

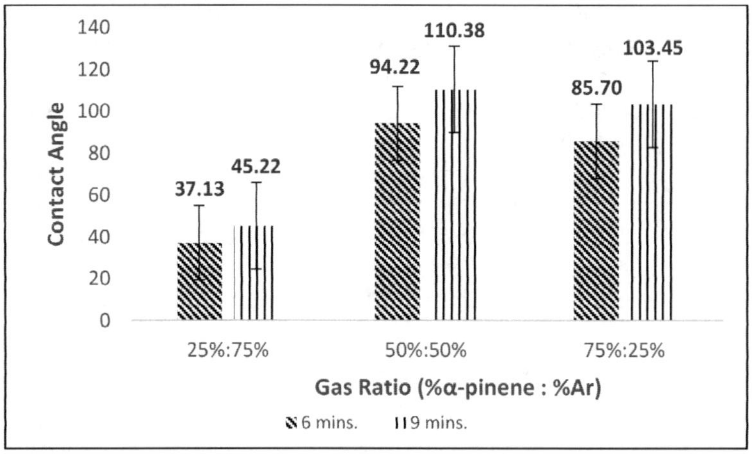

Figure 2. Water contact angle trend for hydrophobicity assessment

FTIR-ATR was done to determine the chemical composition of the deposited α-pinene thin films. Figure 3 shows the FTIR spectral diagram of the plasma polymerized α-pinene of varying monomer proportions. It is observed that several peaks associated with α-pinene were generally no longer present, such as those in 2800-2900 cm^{-1} corresponding to =C-H stretching, while formation of new bands occurred especially at around 2300-2400 cm^{-1}. However, it is shown on Figure 3 that the thin film with 75% α-pinene loading showed low-intensity peaks corresponding to the spectra of the monomer. This is evidenced by the broad and low-intensity peak at 2900 cm^{-1} as well as in 1000 and 1400 cm^{-1}, all suggesting the presence of unpolymerized pinene in the film.

Due to the formation and breaking down of bonds, the plasma polymerized polymer thin film is not considered as a poly(α-pinene) homopolymer, rather it is a polymer derived from α-pinene. The bands of the plasma polymerized pinene were also less well-defined; this implies that that there has been a decrease in the bond homogeneity. This decrease in bond homogeneity can be correlated to the crosslinking that occurs during plasma polymerization; in this process, plasma polymers usually have random bonding and cross-linking. [4,14]

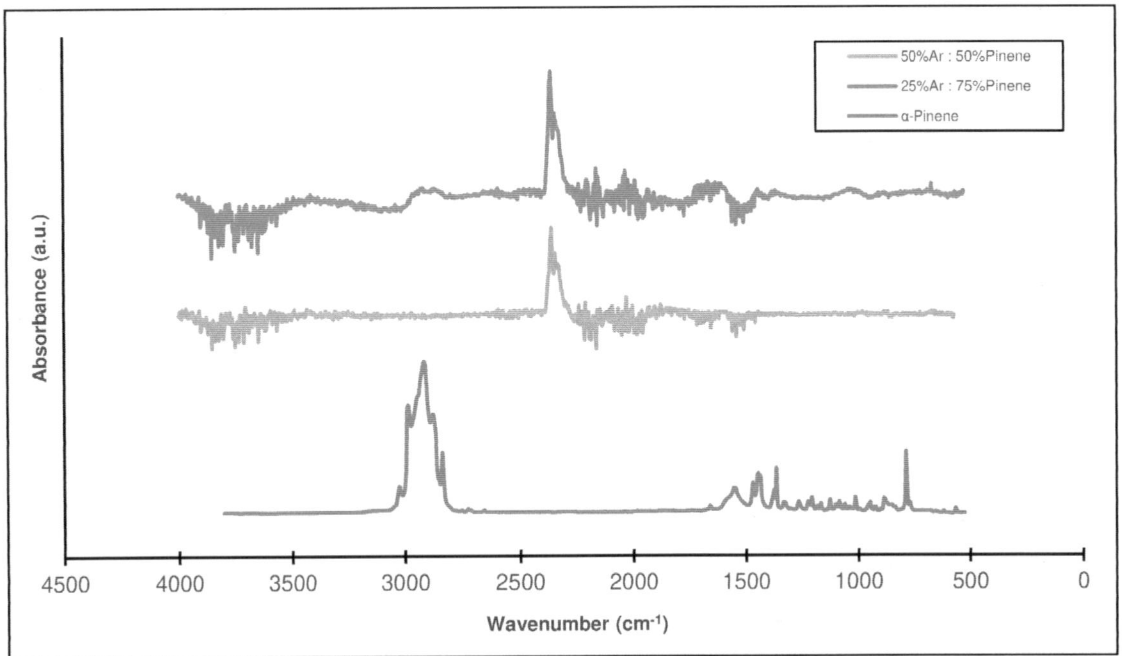

Figure 3. FTIR spectra for the pinene-based thin films with 6.0 minutes deposition time and the indicated gas ratios. FTIR spectra of α-pinene (NIST, 2018) included for comparison

Figure 4 shows the SEM images of the plasma polymerized pinene deposited under 9 mins., with varying gas ratios. It is observed from the figure that for samples deposited with 25:75 and 50:50 gas ratios, no features were apparent, signifying a uniform deposition of the plasma polymerized pinene. A smooth surface finish was observed for the thin film, this implies that the formation of a flat plasma-polymerized pinene was able to level the surface roughness and imperfections of the substrate. However, for the sample with 75:25 gas ratio, bubble-like features were observed, and this is possibly due to the impurities present on the surface of the silicon glass wafers.

Figure 4. FESEM micrographs of the plasma-polymerized pinene thin films deposited on Si substrates taken at 1000x. A pinhole-free image is seen despite varying α-pinene concentrations of 25%, 50%, and 75% (from left to right)

4. Conclusions

A thin film layer of α-pinene was successfully deposited onto the Al substrate. IR analysis suggested that the composition of the thin film only reflects select hydrocarbon bonds and that the plasma polymerization route eliminated the unsaturation present in the α-pinene precursor. On the other hand,

IOP Conf. Series: Materials Science and Engineering **894** (2020) 012009 doi:10.1088/1757-899X/894/1/012009

inspection of the surface morphology of thin film revealed a featureless, pinhole-free surface, consistent with the typical morphology of plasma-polymerized films. For each pinene-Ar gas ratio, it was observed that the hydrophobicity of the Al surface significantly improved as deposition time was increased from 6.0 to 9.0 minutes. The pinene-Ar gas ratio that exhibited the superior hydrophobicity was found out to be at 50:50 gas ratio, regardless of the deposition time, with average water contact angles at 94.22° and 110.38° for samples subjected to 6.0 and 9.0-minute deposition times, respectively. The findings in this work suggest that α-pinene can indeed be deposited as a thin film. This may prove to be significant in the development of organic thin films for a variety of applications such as coating for medical equipment and in the study of organic thin film transistors for flexible electronics applications.

5. References

[1] R. Förch *et al.*, "Recent and Expected Roles of Plasma-Polymerized Films for Biomedical Applications," *Chem. Vap. Depos.*, vol. 13, no. 6-7, pp. 280–294, Jul. 2007.

[2] K. Schröder *et al.*, "Capability of Differently Charged Plasma Polymer Coatings for Control of Tissue Interactions with Titanium Surfaces," *J. Adhes. Sci. Technol.*, vol. 24, no. 7, pp. 1191–1205, Jan. 2010.

[3] C. Rullyani, M. Ramesh, C.-F. Sung, H.-C. Lin, and C.-W. Chu, "Natural polymers for disposable organic thin film transistors," *Org. Electron.*, vol. 54, pp. 154–160, 2018.

[4] D. Gerchman, B. Bones, M. B. Pereira, and A. S. Takimi, "Thin film deposition by plasma polymerization using d-limonene as a renewable precursor," *Prog. Org. Coatings*, vol. 129, pp. 133–139, 2019.

[5] M. Firdaus, L. Montero de Espinosa, and M. A. R. Meier, "Terpene-Based Renewable Monomers and Polymers via Thiol–Ene Additions," *Macromolecules*, vol. 44, no. 18, pp. 7253–7262, Sep. 2011.

[6] B. Rahn, "What Is Pinene and What Are the Benefits of This Cannabis Terpene?," *Leafly*, 2015. .

[7] K. Ložienė, J. Švedienė, A. Paškevičius, V. Raudonienė, O. Sytar, and A. Kosyan, "Influence of plant origin natural α-pinene with different enantiomeric composition on bacteria, yeasts and fungi," *Fitoterapia*, vol. 127, pp. 20–24, 2018.

[8] H. Conrads and M. Schmidt, "Plasma generation and plasma sources," *Plasma Sources Sci. Technol.*, vol. 9, no. 4, pp. 441–454, 2000.

[9] Y.-M. Ko, H.-C. Choe, S.-C. Jung, and B.-H. Kim, "Plasma deposition of a silicone-like layer for the corrosion protection of magnesium," *Prog. Org. Coatings*, vol. 76, no. 12, pp. 1827–1832, 2013.

[10] K. Bazaka and M. V Jacob, "Synthesis of radio frequency plasma polymerized non-synthetic Terpinen-4-ol thin films," *Mater. Lett.*, vol. 63, no. 18, pp. 1594–1597, 2009.

[11] A. Pegalajar-Jurado, C. D. Easton, K. E. Styan, and S. L. McArthur, "Antibacterial activity studies of plasma polymerised cineole films," *J. Mater. Chem. B*, vol. 2, no. 31, pp. 4993–5002, 2014.

[12] Y. W. Chan, K. S. Siow, P. Y. Ng, U. Gires, and B. Yeop Majlis, "Plasma polymerized carvone as an antibacterial and biocompatible coating," *Mater. Sci. Eng. C*, vol. 68, pp. 861–871, 2016.

[13] F. G. Herring, J. D. Madura, C. Bissonette, and R. H. Petrucci, *General Chemistry: Principles and Modern Applications*, 10th ed. Toronto: Pearson Canada, 2011.

[14] D. D. Burkey and K. K. Gleason, "Structure and Thermal Properties of Thin Film Poly(α-methylstyrene) Deposited via Plasma-Enhanced Chemical Vapor Deposition," *Chem. Vap. Depos.*, vol. 9, no. 2, pp. 65–71, Mar. 2003.

Morphology, Thermal Degradation and Dynamic Mechanical Properties of Polyurea used as Explosive-Proof Elastomer

K Won-in[1] and P Dararutana[2]

1 Department of Earth Sciences, Faculty of Science, Kasetsart University, Bangkok, Thailand
2 Retired Army Officer, Royal Thai Army, Bangkok, Thailand

E-mail: kritwonin@gmail.com

Abstract. It is well known that the blast affected the explosion and terrorism, not only human life, but also the building structure. In order to prevent these, protective materials are developed. One of them is polyurea. In this work, the elemental composition and microstructure of the selected polyurea sheet samples are characterized. Thermal, dynamic mechanical, tensile, and hardness properties are also studied. It is found that carbon and oxygen are the main compositions, while silicon, fluorine, and titanium are presented as minor ones. The modulus and loss factor increase as the frequency and temperature increase. The elongation property is over 100% and the hardness is around 90 Shore A.

1. Introduction

Because of the destroying property of the explosion, it is manifested that the blast and fragmentation can affect the injury and death [1-3]. It is revealed that some elastomeric polymers can improve the coated structure from the blast of the explosion. Polyurea is one of them, which consists of both soft and hard regions [4-8]. It is applied for civilian and military purposes.

It is well known that the dynamic mechanical properties of materials consisted of storage modulus (E'), loss modulus (E'') and loss factor (tanδ). These values considered the amount energy stored as the elastic energy and the energy dissipated during the straining process. The E' is defined as evaluated the elastic property of the materials and represented the modulus in the bending of specimens. It decreases rapidly and then reaches a plateau, according to the temperature increased. Until at higher temperatures, the elongation rubber modulus maintains a constant value and independent with the frequencies. The E'' which is exhibited the amount of the energy absorbed by the materials under the alternating stress action indicated the viscous performance of the viscoelastic materials. It increases with the temperature increased to the maximum value around the glass transition temperature (T_g) and then decreased rapidly affected by the change in the intramolecular friction. The tanδ is defined as the ratio between the E'' and E' and where the δ equaled the phase angle between the stress and the strain. The peaks of the tanδ increased as the frequency increased. It is also revealed by other authors that the changes of these dynamic mechanical properties, such as E', E'', and tanδ are varied with the temperature and frequency increase [9-17].

This work, the microstructure and elemental composition of polyurea samples are studied using a scanning electron microscope coupled with an energy dispersive X-ray spectrometer (SEM-EDS). Thermogravimetric (TG), dynamic mechanical analysis (DMA), universal testing machine (UTM), and Shore A durometer are carried out to test the thermal, dynamic mechanical, tensile, and hardness properties, respectively.

2. Experimental

2.1 Sample

Five samples of polyurea sheets were selected from various products of the Seashore Resources Pte.Ltd. (Singapore), which made by the spray coating technology. They were the difference in colorations, such as dark-gray, gray, light gray, red-brown and smoke gray, labeled as PU-1, PU-2, PU-3, PU-4, and PU-5, with the thickness of 0.43, 2.08, 1.60, 1.82 and 1.07 mm, respectively, as shown in figure 1.

PU-1 PU-2 PU-3 PU-4 PU-5

Figure 1. The polyurea sheet samples

2.2 Methods

2.2.1 SEM-EDS. The microstructure and elemental composition of the samples were analyzed using a scanning electron microscope (SEM) QUANTA 450 (FEI Company, USA) coupled with an energy dispersive X-ray spectrometer (EDS) X-Max (Oxford Instruments, UK) at the Science Equipment Center of Faculty of Science (Kasetsart University, Thailand). The SEM system was operated at 15 kV, and magnification of 10,000X and 30,000X. Quantitative analysis of the XRF spectra was performed using the INCA program (Oxford Instruments, UK). Prior to characterize, their surface was coated with gold using Sputter coater SC7620 (Quorum, UK). For each sample, X-ray spectra were reported at three different positions.

2.2.2 Thermal degradation. The thermal degradation of the samples was analyzed by a TGA2 (Mettler Toledo, Switzerland) at the Mettler-Toledo (Thailand) Limited (Bangkok, Thailand), with approximately 9.50± to 1.00 mg. sample weight, under the nitrogen atmosphere at a heating rate of 10ºC/min. from 25 to 700ºC. For each sample were conducted three records to assure acceptable results.

2.2.3 Dynamic property. The dynamic properties of the samples were carried out with a dynamic mechanical analyzer (DMA), DMA1 (Mettler Toledo, Switzerland) at the Mettler-Toledo (Thailand) Limited (Bangkok, Thailand). Measurements were designed by applying a temperature range from -100 to 140ºC at the constant rate of 10ºC/min., while the frequencies were changed from 1, 3, and 5 Hz. Three records for each sample were analyzed.

2.2.4 Hardness. The ASTM standard 2240, ISO/R868 was carried out to test the hardness of the samples at the Faculty of Science of Chiang Mai University (Chiang Mai, Thailand), with a Cone 35 of 1.3 mm in diameter, 0 – 2.5 mm depth of indentation and 250 g gross weight.

2.2.5 Tensile property. A universal testing machine (UTM) was used to test the tensile properties of the samples, prepared with the standard shape specimens at the Faculty of Engineering of Chulalongkorn University (Bangkok, Thailand).

3. Results and discussions

The result using EDS, as shown in Table 1, showed the elemental composition of the polyurea sheet samples composed of C and O that were as the main composition, while Si, F, and Ti seemed to be the minor ones. It was shown the presence of C and O were approximately 62.7-74.9 and 17.1-28.7 wt%, respectively. Si was found between 0.2 to 1.7 wt%, while Ti which detected in sample PU-1, PU-2,

PU-3, and PU-5 was the only element that was added to modify the polyurea structure [18]. Fe which only found in sample PU-4 and PU-5 was added as the pigment to show reddish-brown coloration. S and Cl were found in sample PU-1 which were introduced as the chain extenders; dimethylthio-toluenediamine (DMTDA) and 4, 4'-metylenebis-(3-chloro, 2, 6-diethyl)-aniline (MCDEA), respectively. Other elements such as F, Na, Mg, Al, Ca, and K which showed in some samples were added as additives, fillers, and pigments.

The SEM analysis showed the microstructure which consisted of both partially and fully soften regions, that were likewise the other author [19]. They showed the nanometer size hard domains, as shown in figure 2. It was known that polyurea which was mixed with micro- and nano-particles resulted to modify its properties [8]. The hard domains in polyurea were reported that they acted as both inter-chain links and nanometer-size reinforcements making a nano-scale composite material with a unique set of mechanical properties [19].

Table 1. The elemental composition of polyurea sheet samples

Sample	Elemental composition (wt%)												
	C	O	F	Na	Mg	Al	Si	S	Cl	Ca	K	Ti	Fe
PU-1	62.7	28.7	nd	0.6	0.3	0.4	1.7	0.1	0.4	0.2	0.3	4.5	nd
PU-2	73.1	20.4	4.4	0.2	nd	0.2	0.7	nd	nd	nd	0.2	0.5	nd
PU-3	74.9	24.4	nd	nd	nd	nd	0.2	nd	nd	nd	nd	0.6	nd
PU-4	73.0	17.1	9.2	nd	nd	0.1	0.4	nd	nd	nd	0.1	nd	0.2
PU-5	74.8	24.0	nd	nd	0.2	nd	0.6	nd	nd	nd	nd	0.5	0.1

Note: nd equals to non-detectable; less than 0.1 wt

The thermal behavior of the samples was evaluated by the TGA and the thermal gravimetrical trace was shown in figure 3. The thermogram of PU-1 exhibited a single-step decomposition profile, while others showed more than ones.

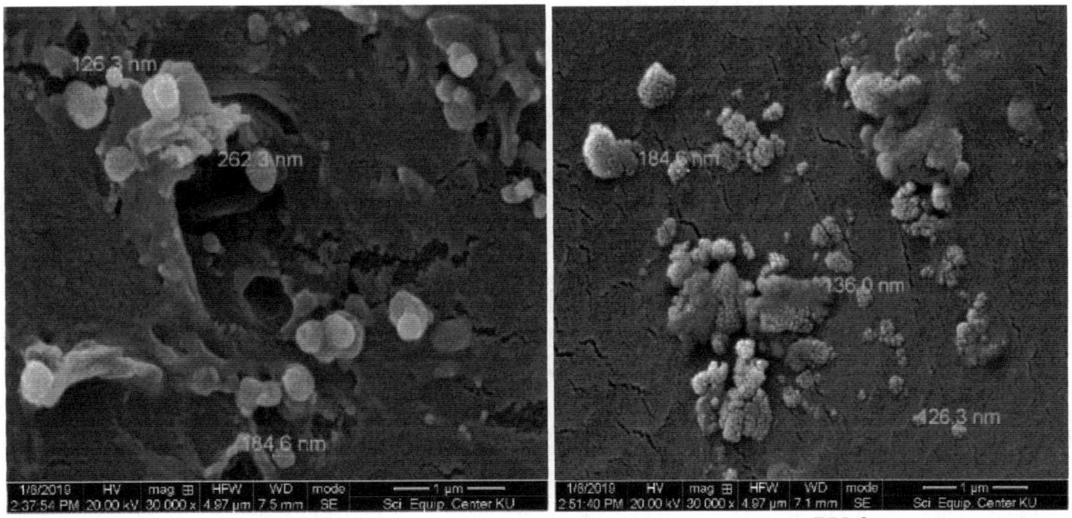

PU-1 PU-2

IOP Conf. Series: Materials Science and Engineering **894** (2020) 012010 doi:10.1088/1757-899X/894/1/012010

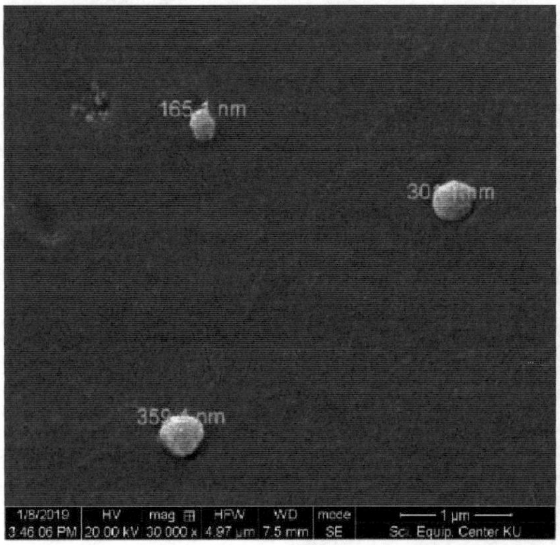

PU-3

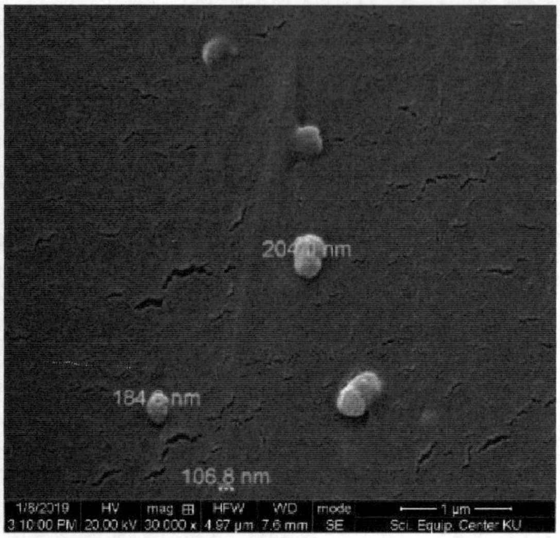

PU-4

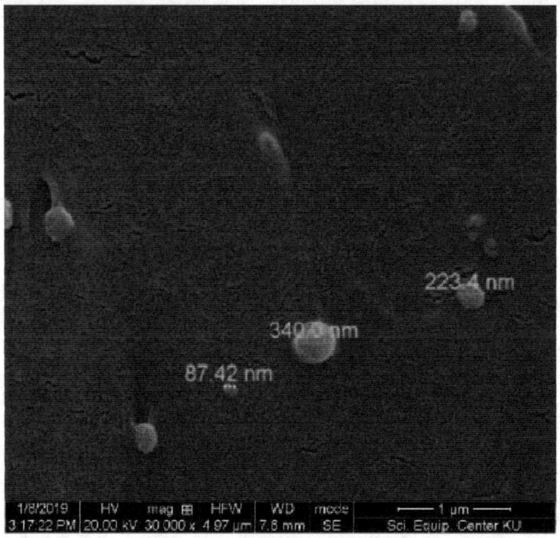

PU-5

Figure 2. The SEM micrographs of the polyurea sheet samples with magnification at 30,000X

The obtained values of the *E'*, *E''*, and T_g resulted in the various frequencies, such as 1, 3, and 5 Hz, as presented in table 2 and figure 4. It was shown that the *E'*, *E''*, and T_g values were mostly in agreement with the previous works, except PU-5 which these values differed from them. It was found that the higher the T_g was, the higher the frequency was.

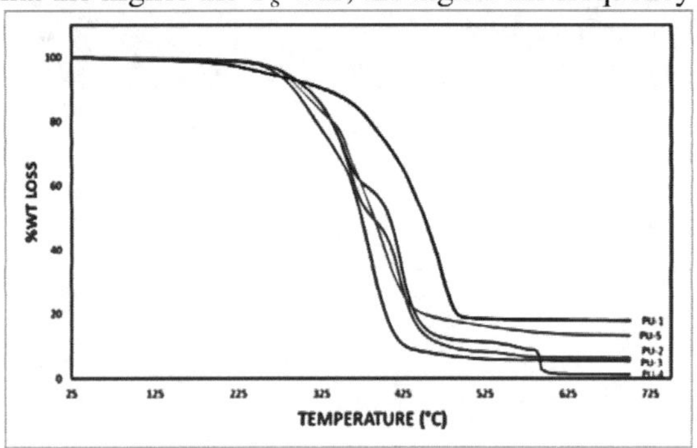

(a) TGA

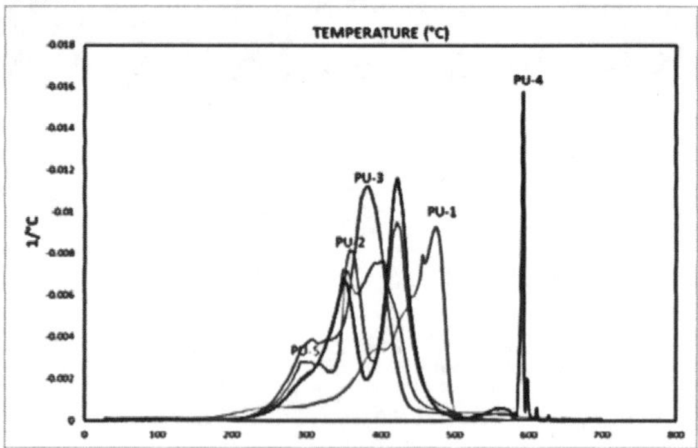

(b) DTA

Figure 3. The thermal degradation tests for polyurea sheet samples.

Table 2. The DMA data of polyurea sheet samples.

Sample	Storage modulus, *E′* (Mpa)			Loss modulus, *E″*(Mpa)			Glass transition temperature, T_g (°C)		
	1 Hz	3 Hz	5 Hz	1 Hz	3 Hz	5 Hz	1 Hz	3 Hz	5 Hz
PU-1	3910	4265	4220	142.4	143.4	143.6	68.02	73.88	79.39
PU-2	2530	2600	2650	6.8	8.1	8.5	-10.02	-11.89	-8.63
PU-3	3820	3790	3800	237.2	237.1	237.0	-16.00	-9.97	-6.73
PU-4	2350	2370	2390	125.0	120.4	119.5	-19.77	-16.33	-15.67
PU-5	2040	2010	2050	135.5	132.9	133.7	-15.80	-17.62	-17.05

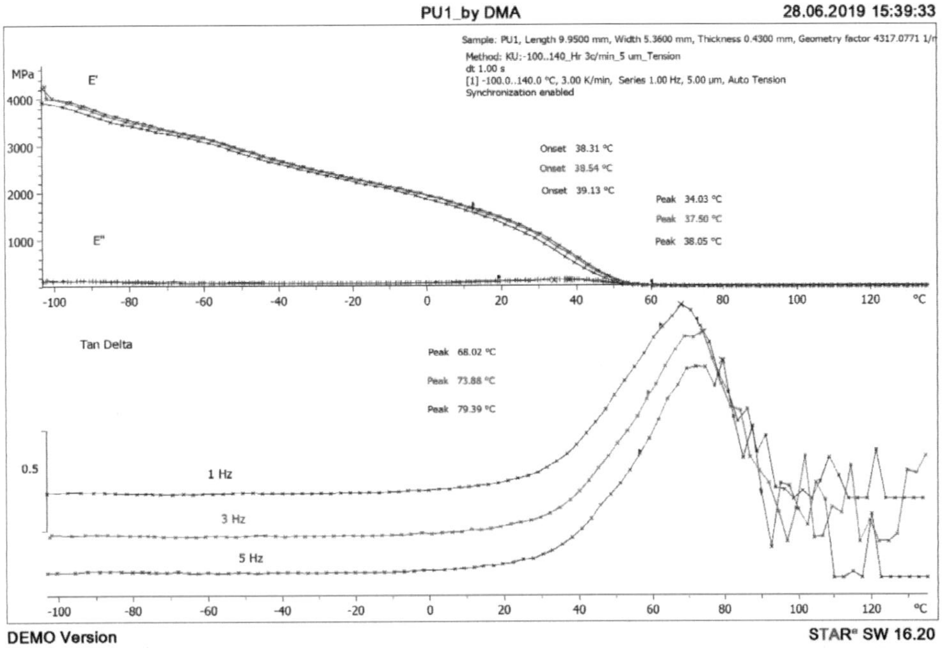

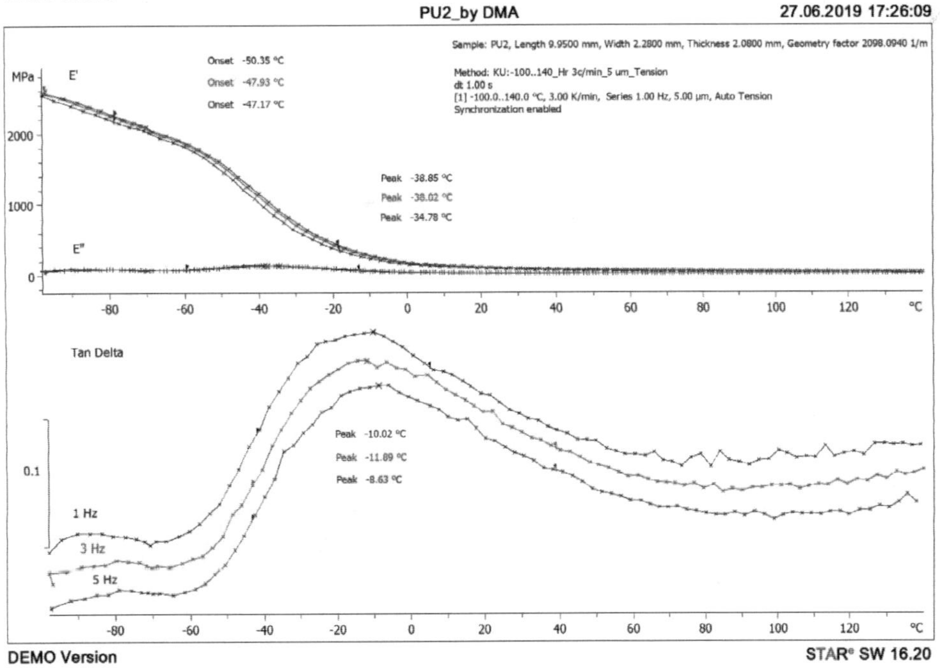

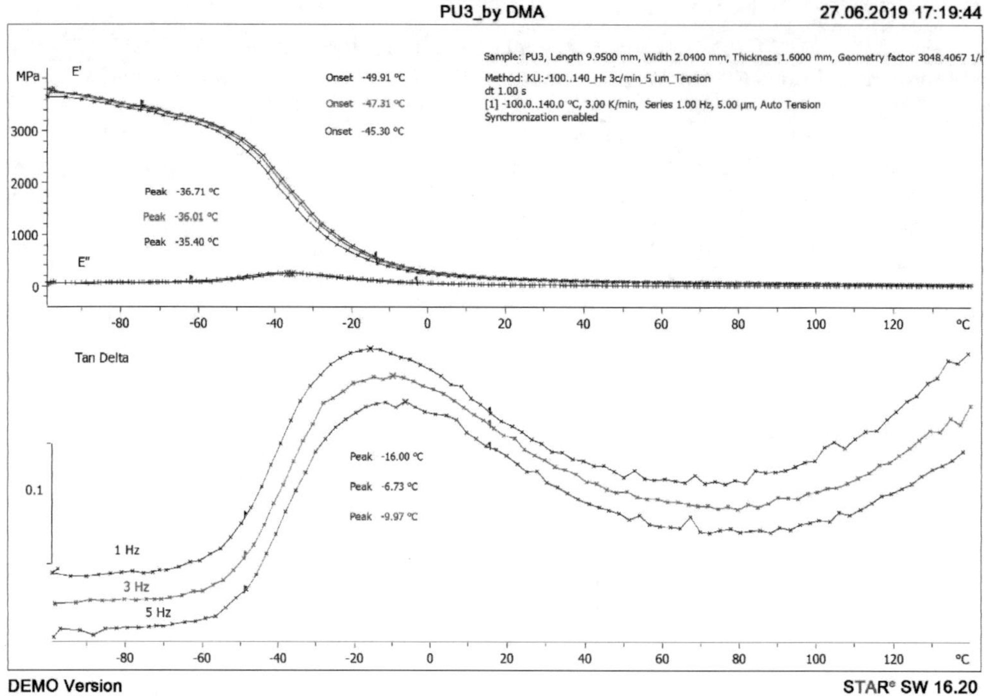
PU3_by DMA
27.06.2019 17:19:44
Sample: PU3, Length 9.9500 mm, Width 2.0400 mm, Thickness 1.6000 mm, Geometry factor 3048.4067 1/
Method: KU:-100..140_Hr 3c/min_5 um_Tension
dt 1.00 s
[1] -100.0..140.0 °C, 3.00 K/min, Series 1.00 Hz, 5.00 µm, Auto Tension
Synchronization enabled
MPa
E'
E"
Onset -49.91 °C
Onset -47.31 °C
Onset -45.30 °C
Peak -36.71 °C
Peak -36.01 °C
Peak -35.40 °C
Tan Delta
Peak -16.00 °C
Peak -6.73 °C
Peak -9.97 °C
0.1
1 Hz
3 Hz
5 Hz
°C
DEMO Version
STAR® SW 16.20

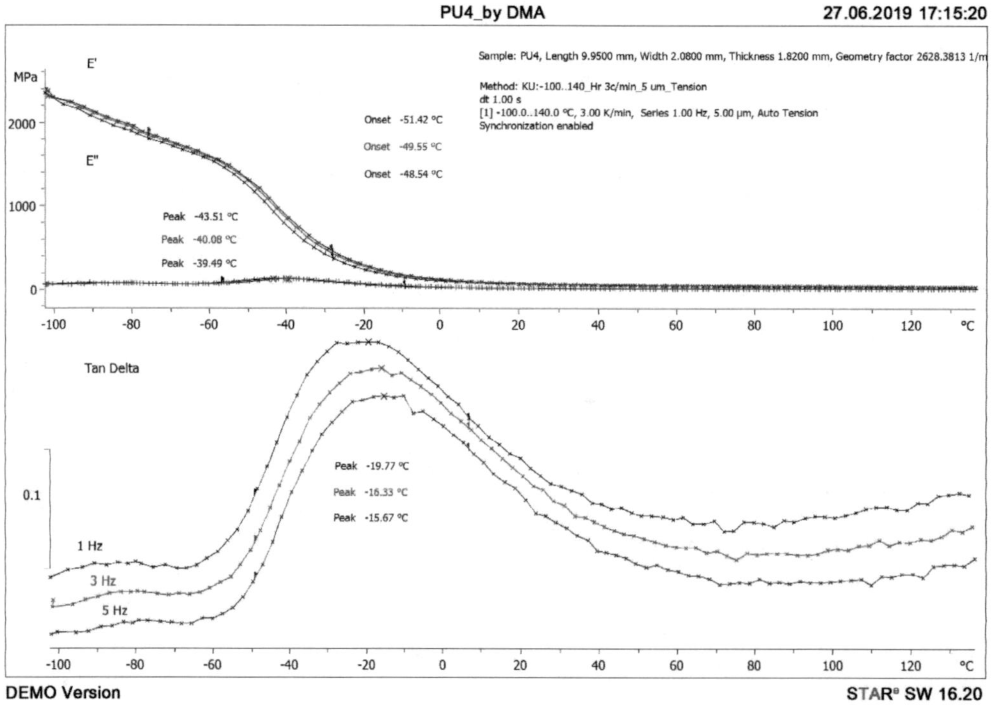
PU4_by DMA
27.06.2019 17:15:20
Sample: PU4, Length 9.9500 mm, Width 2.0800 mm, Thickness 1.8200 mm, Geometry factor 2628.3813 1/m
Method: KU:-100..140_Hr 3c/min_5 um_Tension
dt 1.00 s
[1] -100.0..140.0 °C, 3.00 K/min, Series 1.00 Hz, 5.00 µm, Auto Tension
Synchronization enabled
MPa
E'
E"
Onset -51.42 °C
Onset -49.55 °C
Onset -48.54 °C
Peak -43.51 °C
Peak -40.08 °C
Peak -39.49 °C
Tan Delta
Peak -19.77 °C
Peak -16.33 °C
Peak -15.67 °C
0.1
1 Hz
3 Hz
5 Hz
°C
DEMO Version
STAR® SW 16.20

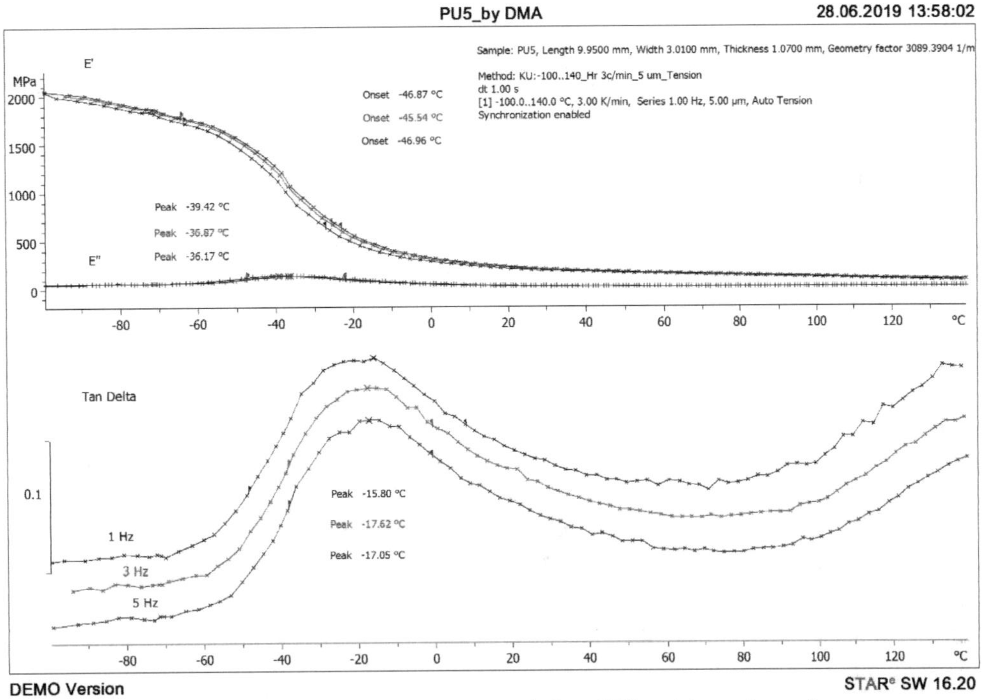

Figure 4. The storage modulus (E') and loss modulus (E'') with various frequency curves of polyurea sheet samples.

It was also found that the tensile strength, modulus, and elongation at break were varied approximately between 7 - 20 MPa, 500 - 770 and 100 - 440%, respectively, as shown in table 3.

The values of the hardness at different points were also found between 87.0-99.0 Shore A, as illustrated in table 4.

Table 3. The tensile property of the polyurea sheet samples.

Sample	**Tensile property**		
	Tensile strength (MPa)	**Elongation at Break (%)**	**Modulus (MPa)**
PU-1	18	260	735
PU-2	20	440	770
PU-3	15	375	710
PU-4	7	100	500
PU-5	19	500	550

Table 4. The hardness property of the polyurea sheet samples.

Sample	**Hardness, Shore A**
PU-1	99.0
PU-2	87.0
PU-3	94.0
PU-4	87.0
PU-5	90.0

4. Conclusion

Polyurea sheet samples were successfully characterized, including microstructure, dynamic mechanical behavior, thermal stabilities, tensile and hardness properties.

They were analyzed by SEM-EDS, in order to characterize the materials used and obtain the information on their production technology, especially the elemental addition, such as Ti and Fe for the structural modification and pigment, respectively. The surface morphology using SEM showed amorphous nature which contained both the micro- and nano-particles. The EDS results showed the presence of C, O, and Si in all samples. Other elements were added as the chain extenders, additives, fillers, and pigments.

Using TGA, DMA, UTM, and Durometer methods were possible to demonstrate the thermal, tensile, and hardness properties of the polyurea sheet samples, respectively.

Acknowledgment
The work has been facilitated within the activity of the Kasetsart University (Department of Earth Science at the Faculty of Science and the Science Equipment Center). The Sciences Equipment Center at Faculty of Science of Kasetsart University, Mettler-Toledo (Thailand) Limited, LMS Instruments Co.,Ltd., the Faculty of Engineering at Chulalongkorn University, and the Faculty of Science at Chiang Mai University thank for microstructure and elemental composition analysis, DMA and TGA, UTM, and hardness measurements, respectively.

5.References

[1] Raman S N, Ngo T and Mendis P 2011 *Journal of Structural Engineering* **11** 69.
[2] Raman S N, Ngo T, Mendis P and Pham T 2012 *Advances in Materials Science and Engineering* ID754142(2012)008.
[3] Dai L-H, Wu C, An F-J and Liao S-S 2018 *Advances in Materials Science and Engineering* ID1264276(2012)007.
[4] Grujicic M, Bell W C, Pandurangan B and He T 2010 *Materials and Design* **31** 4050.
[5] Mariappan T and Wilkie C A 2013 *Polimery* **58** 371.
[6] Toutanji H A, Choi H, Wong D, Gilbert J A and Alldredge D J 2013 *Construction and Building Materials* **38** 1170.
[7] Qiao J, Amirkhizi A V and Nasser S-N 2015 *Proc. of SPIE 9432* (San Diego: USA) 94320R.
[8] Zhai W, He J, Li X, Chen Q, Zhen J and Wang X 2018 *IOP Conf. Series: Materials Science and Engineering* 423(2018)006.
[9] Huang H, Yang Y, Lyu P and Zhang R 2017 *International Conference on Transportation Infrastructure and Materials,* June 9-12, 2017 (Qingdao: China) 10 pp.
[10] Das S, Yilgor I, Yilgor E, Inci B, Tezgel O, Beyer F L and Wilkes G L 2007 *Polymer* **48(1)** 290.
[11] Lv P, Chen G and Huang W B 2007 *Journal of Wuhan University of Technology* **29(4)** 61.
[12] Vlad S and Oprea S 2007 *Journal of Optoelectronics and Advanced Materials* **9(4)** 994.
[13] Pathak J, Twigg J, Nugent K, Ho D, Lin E, Mott P, Robertson C, Vukmir M, Iii T E and Roland C 2008 *Macromolecules* **41(20)** 7543.
[14] Fragiadakis D, Gamache R, Bogoslovov R and Roland C 2010 *Polymer* **51(1)** 178.
[15] Huang W B and Lv P 210 *Polymer Materials Science and Engineering* **26(3)** 72.
[16] Gair Jr J L, Lambeth R H, Cole D P, Lidston D L, Stein I Y, Cohen E-K, Hsieh A J, Bruck H A, Bundy M L, Wardle B L *Composites Science and Technology* **in press**.
[17] Jiang M, Liu Q, Zhang Q, Ye C and Zhou G 2012 *Journal of Polymer Science Part A: Polymer Chemistry* **50(5)** 1026.
[18] Russu R and Thomas E 1983 *Journal of Macromolecular Science Part B* **22** 553.
[19] Arunkumar T and Ramachandran S 2015 *Applied Mechanics and Materials* **766-767** 606.

IOP Conf. Series: Materials Science and Engineering **894** (2020) 012011 doi:10.1088/1757-899X/894/1/012011

Role of Organic Additives in the Fabrication of Micro Copper Rods by Localized Electrochemical Deposition

N Alshwawreh[1], O Hussien[1] and M Khanfar[2]

[1]Industrial Engineering Department, School of Applied Technical Sciences, German Jordanian University, Amman Madaba Road, Amman, 11180, Jordan
[2]Pharmaceutical and Chemical Engineering Department, School of Applied Medical Sciences, German Jordanian University, Amman Madaba Road, Amman, 11180, Jordan

Email: Nidal.Alshwawreh@gju.edu.jo

Abstract. In this research, an automated experimental setup to fabricate high aspect ratio copper micro rods using localized electrodeposition was assembled. Using a custom-made microelectrode, copper micro columns were successfully fabricated on copper substrates from an additive-free electrolyte. To expand the capabilities of localized electrochemical deposition as a potential microfabrication technique, the effect of organic additives including Janus Green B (JGB), Polyethylene Glycol (PEG), 3-mercapto-1-propane sulfonic acid (MPSA) on the uniformity of the fabricated structures was studied. Scanning electron microscopy characterization revealed that organic additives play an important role in achieving a uniform and rigid micro rod across its height. An electrolyte containing all three additives (PEG, JGB, MPSA) appears to produce a micro rod with better structural characteristics compared with the case when only two additives are present (PEG, JGB). From this research, designing an electrolyte with proper chemistry appears to be critical to realize optimized fabrication of micro scale features by localized electrodeposition. It could also contribute to presenting the technique as a competitive and reliable micro fabrication alternative to other complex and high-cost technologies.

1. Introduction

Micro-scale and nano-scale structures have enormous applications in advanced systems such as drug delivery in biomedicine, micro-electro-mechanical systems (MEMS), small volume reactors (e.g. micro-analytical devices), and very large integrated circuits in microelectronic chips. Traditionally, these miniaturized systems are fabricated on silicon substrates using processes that often include microlithography, etching, and planarization. The capabilities of these masked-processes are limited to planar thin films structures and in order to produce 3D features, the process has to be repeated many times using one or several masks. One field that is experiencing a rapid growth in advanced 3D machining is the development of additive manufacturing technologies in which the desired object is built layer by layer. Recently, a new method to fabricate microstructures by localized electrochemical deposition (LED) was proposed [1-7]. Compared to other techniques, LED is considered a low-cost additive manufacturing process that does not require a mask. As a consequence, LED is a much simpler tool lending itself to the potential use as a stand-alone microfabrication technique. While other technologies have material-related limitations, LED is able to fabricate structures from a wide range of materials and on different substrates.

The principle of localized electrodeposition technique is based on using a sharp microelectrode (anode) to fabricate structures on a conductive substrate (cathode) by inducing a chemical reaction. In an electrochemical cell, the chemical reaction (oxidation/reduction) takes place when an external electrical potential is applied between the microelectrode tip (which is exposed to the electrolyte) and the substrate. For example, in the presence of an electrolyte containing copper ions, localized deposition of copper will occur underneath the microelectrode tip due to intense electric field at the gap between the tip and the substrate. When the microelectrode is withdrawn away from the substrate surface, the deposit will grow in size and, in this way, high aspect ratio structures can be realized. Although LED technique is relatively simple, the technology is still under development. All research reported in literature include attempts to demonstrate the capabilities of the technique using in-house and laboratory-scale experiments. There is a consensus in the scientific community, however, that in order to adopt LED as a standard microfabrication tool, several technical issues should be resolved which include aspects related to resolution and automation [8]. The aim of this paper is to investigate the effect of electrolyte chemistry and the presence of organic additives on the integrity of copper micro rods fabricated by an automated localized electrodeposition setup. These micro rods with excellent mechanical properties can be useful in many applications like sensors and microelectromechanical devices.

2. Materials and Experimental Procedure

The microelectrode that was used in this work was fabricated in-house. It involves the use of a conductive wire made of platinum/iridium alloy (Goodfellow, United Kingdom). The wire (50-75 μm in diameter in this work) was embedded in a glass matrix and then polished to expose the wire tip at the microelectrode end (see figure 1). To facilitate microelectrode handling, the shank of the microelectrode was supported by Teflon tube. The other end of the microelectrode was soldered with a copper rod to provide necessary electrical contact. The substrate used in this work was made of copper and designed to fit in the center of an electrochemical cell made from Teflon. To control microelectrode trajectory, it was attached to a closed loop piezoelectric actuator with a resolution of about 30 nm (Newport Inc., USA, model CL-8310). The electrical potential to be applied between the microelectrode tip and the polished substrate (3.8 V in this case) was supplied via a data acquisition unit (National Instruments Inc., USA, Model MyRio 1900). The device can be controlled to supply DC voltage with a maximum current of 100 mA (with the aid of a voltage follower). The movement of the piezoelectric motor was controlled by LabVIEW with the aid of the actuator driver.

(a) **(b)**

Figure 1: (a) the assembled localized electrodeposition setup (b) the tip of the fabricated microelectrode (50 μm tip diameter, 600μm glass diameter around the tip).

IOP Conf. Series: Materials Science and Engineering **894** (2020) 012011 doi:10.1088/1757-899X/894/1/012011

The base electrolyte that was used in this research consists of 0.4M $CuSO_4.5H_2O$ and 1.8M H_2SO_4. When the voltage was applied between the tip and the substrate, the electrochemical current resulting from the oxidation/reduction reaction was monitored by LabVIEW via the data acquisition unit. In this way, the current was employed as a feedback to assess the rate of electrochemical deposition taking place underneath the microelectrode tip. To ensure successful deposition, the initial positioning of the microelectrode relative to the substrate was controlled such that a gap less than 100 μm is maintained. The system was designed such that when the deposition current reaches a value greater than 10 mA (an indication that the deposited copper has filled the gap underneath the tip), the LabVIEW code triggers a signal to withdraw the piezoelectric motor away from the substrate at a constant speed by a certain distance. This allows the copper deposit to continue growing underneath the microelectrode tip. The process was repeated until the desired height of the micro rod was achieved. To investigate the effect of organic additives, Janus Green B (JGB, 1mg/L) and polyethylene glycol (PEG, 300 mg/L) were added to the electrolyte. In addition, and in order to improve the effect of PEG, potassium chloride (KCl, 105mg/L) was added. Another experiment includes adding 3-mercapto-1-propane sulfonic acid (MPSA 15mg/L) in the presence of other additives. These additives are known to improve the characteristics of copper deposit and is normally used in microelectronics industry to fabricate copper interconnects and thin films in very large integrated circuits and microprocessors. After finishing the deposition experiment, the substrate with the fabricated micro rod was removed from the electrochemical cell, rinsed in deionized water and then examined by scanning electron microscopy (TESCAN VEGA) at 25 kV to analyse the effect of organic additives on the structure characteristics. In order to facilitate imaging, the substrate was tilted by rotating the microscope stage to capture the entire geometry of the fabricated structure.

3. Results and Discussion

Figure 2 shows examples of fabricated copper structures by localized electrodeposition in the absence of organic additives in the electrolyte. As shown in figure 2-a, localized electrodeposition was successful in the fabrication of copper micro rod that is several millimetres in height. However, a magnified version of this high aspect ratio structure (see figure 2-b) shows that the lateral dimension is not consistent along its length. Moreover, the top part of the rod appears to be mechanically weak as evident from its bent tip. These irregularities in the shape and dimensions are not desired and may deteriorate the performance of these structures when integrated in real engineering application. To ensure repeatability, another experiment under the same conditions was conducted (see figure 2-c and 2-d). Similar irregularities were observed which appear to initiate at the early stages of the deposition. While the second micro rod appears to be straight till the tip, the rod appears to be rough at the base with some cracks or openings that are not completely filled with copper.

The addition of organic additives to the electrolyte appears to have a significant effect on the morphology and consistency of the deposited micro column. This is illustrated by the SEM images presented by figure 3. Here, the addition of PEG and JGB to the electrolyte resulted in a successful deposition of micro rods with several millimetres in height (similar to the result obtained from additive-free electrolyte experiments). However, the main difference is that the diameter of the rod from the base to the top of the structure appears to be uniform (average diameter of about 51±2 μm). This indicates that the rod diameter almost exactly matches the diameter of the tip (in this case, the tip diameter was 50 μm). While the structure appears to be thicker than the structures fabricated from additive-free electrolyte, the rod is still porous and these defects are present across its height. It is believed that this may be due to the presence of PEG as a suppressor for copper deposition. Adding MPSA to the electrolyte with the other two organic additives results in a denser micro rod as illustrated in figure 4. In addition to the successful realization of high aspect ratio structure, the density of pores at the surface appears to be less compared with the case when MPSA is absent. However, a closer look at the rod surface at higher magnification reveals relatively rough surface. Here, the structure appears to be composed of larger copper grains. This may indicate that MPSA or its interaction with the other two additives appears to limit the nucleation rate of copper crystals during electrodeposition of micro rod structures.

IOP Conf. Series: Materials Science and Engineering **894** (2020) 012011 doi:10.1088/1757-899X/894/1/012011

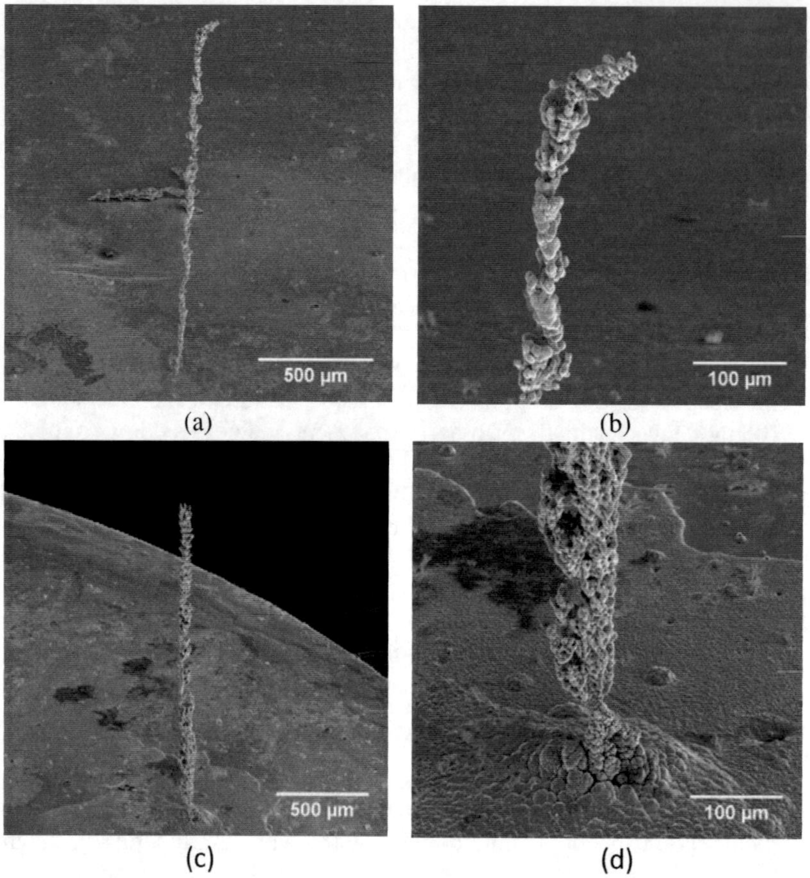

Figure 2: SEM images for copper structures fabricated by localized electrodeposition using an additive-free electrolyte.

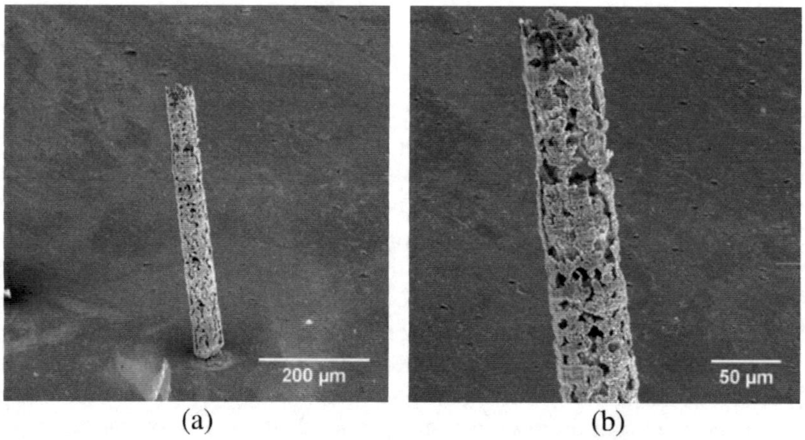

Figure 3: SEM images of micro rods that were deposited using an additive-containing electrolyte (PEG, JGB) (a) whole micro rod structure (b) higher magnification of the top part of the structure.

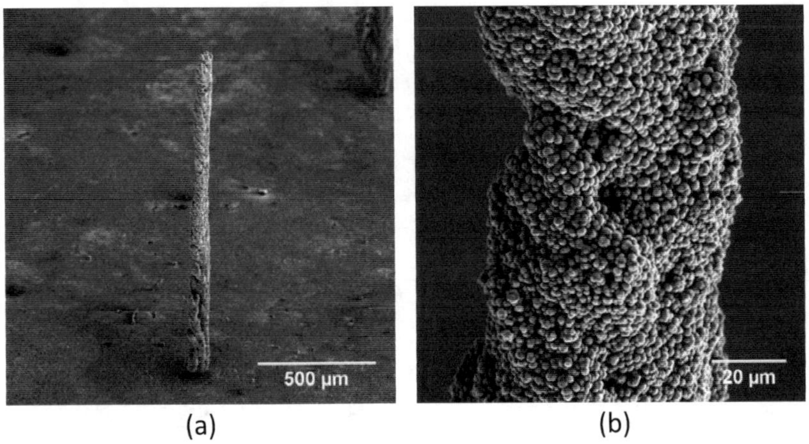

Figure 4: SEM images of a micro rod that was deposited using an additive-containing electrolyte (PEG, JGB, and MPSA) (a) the whole micro rod (b) the surface of the rod at higher magnification.

Normally, organic additives are utilized in the fabrication of copper interconnects by electrodeposition. The superfilling of deep structures by this technique was explained by the inhibition effect of PEG, the acceleration effect of MPSA, and with the leveling effect of JGB and their different diffusion kinetics. Moffat et al. [9] proposed a theory based on curvature-enhanced accelerator coverage. In this mechanism, there is a competition between inhibitors and accelerators and the accelerator is assumed to remain at the surface during the deposition. In the research presented here, there is no sidewalls and the explanation of the observed improvement in the fabrication of micro rods by organic additives can be better described by curvature-enhanced accelerated coverage theory. However, in-depth experimental studies are needed to confirm this with proper measurement of organic molecules distribution in the deposited structure. In addition, the concentration of the organic additives should be further refined to produce a more homogenous and smooth surface.

4. Conclusion

High aspect ratio copper micro rods were successfully fabricated by localized electrodeposition setup controlled by LabVIEW. Organic additives in the electrolyte like JGB, PEG, and MPSA appear to improve the characteristics of the deposited structure. Micro rods fabricated with the presence of MPSA in the electrolyte appear to be of a better quality than the case when MPSA is absent. However, the concentration of organic additives in the bath and its effect on the smoothness of the surface and mechanical properties is yet to be investigated.

Acknowledgement

This project was funded by Scientific Research and Innovation Fund, Ministry of Higher Education and Scientific Research of Jordan (Fund#: ENG/1/29/2017).

5. References

[1] Madden J and Hunter I 1996 *J. Microelectrochemical systems* **5** 24-32
[2] El-Giar E, Said R, Bridges G and Thomson D 2000 *J. Electrochem. Soc.* **147** 586-91
[3] Jansson A, Thornell G and Johansson S 2000 *J. Electrochem. Soc.* **147** 1810-17
[4] Lin J, Chang T, Yang J, Chen Y and Chuang C 2010 *Electrochimica Acta* **55** 1888-94
[5] Lin C, Lee C, Yang J and Huang Y 2005 *Electrochem. Solid State Lett.* **8** C125-9
[6] Yang J, Lin J, Chang T, Lai G and Jiang S 2003 *J. Micromech. Microeng.* **18** 055023
[7] Volgin V, Lyubimov V, Gnidina I, Davydov A and Kabanova T 2018 *Procedia CIRP* **68** 242-7
[8] Said R 2003 *Nanotechnology* **14** 523-31
[9] Moffat T, Wheeler D and Josell D 2004 *Electrochem. Soc. Interface* **13** 46

IOP Conf. Series: Materials Science and Engineering **894** (2020) 012012 doi:10.1088/1757-899X/894/1/012012

Study of Forming Efficiency by Real-time Correction of Head Nozzle Height in Directed Energy Deposition

Kengo Aizawa[1, a], **Masahiro Ueda**[1, 2, b], **Hideki Aoyama**[1, c], **Kazuo Yamazaki**[1, 2, 3, d]

[1] Keio University, 3-14-1 Hiyoshi, Kohoku-ku, Yokohama, 223-8522 Japan
[2] University of California, Berkeley, CA 94720-1740
[3] Bainer Hall, University of California, Davis, CA 95616-5294

[a]aizawa_k@ddm.sd.keio.ac.jp, [b]m-ueda@berkeley.edu, [c]haoyama@sd.keio.ac.jp, [d]kyamazaki@ucdavis.edu

Abstract. Laser metal deposition is of two types: powder bed and metal deposition. Either method is expected to be a breakthrough in innovative design and manufacturing technology development because both methods can form shapes that cannot be formed by conventional removal processing. One metal deposition method is directed energy deposition (DED), in which a laser irradiates the target position on the metal layer to melt it, and metal powder is supplied to the target position to perform molding. DED is expected to produce large parts and repair molds. However, as the lamination mechanism is not sufficiently clarified, relationships between process parameters, material properties, molding accuracy, and molding efficiency are not clear, making stable molding difficult [1]One obstacle to stable molding is that the distance between the head nozzle that supplies the metal powder and the molten pool cannot currently be kept constant. In this study, we measure the distance between the head nozzle and the molten pool using a triangulation method, propose a method to control the position of the head nozzle to keep the distance constant, and experimentally verify the method's usefulness.

1. Introduction

In additive manufacturing (AM), design and manufacture can be used continuously and integrally based on three-dimensional (3D) data, because the object is formed using digital data of the object structure created by 3D computer-assisted design (CAD) or computer graphics (CG). As a result, it is possible in principle to produce a target object while minimizing both the amount of materials and the number of processes, and in recent years this approach has attracted attention as a way to manufacture at low cost and with low environmental impact. AM encompasses several methods. Among them, the method of using a laser and a metal material is generically referred to as laser metal deposition. One metal deposition method is directed energy deposition (DED), which attaches a laser head device and a powder supply device to a machine tool and performs lamination. DED is a multidisciplinary technology that uses laser technology, CAD/CAM, robotics, sensors, controls, and powder metallurgy. However, the final characteristics of parts fabricated by DED depend on various process parameters such as laser power, powder mass flow rate, scanning rate, carrier gas flow rate, and shield gas flow rate. Therefore, many studies have focused on technical performance such as hardness, wear resistance, phase analysis, and corrosion resistance of parts molded using DED [2],[3],[4],[5].

In order to obtain the desired geometrical dimensions and material properties in a finished product, it is important to control the process parameters. Many researchers are currently examining the relationship between process parameters and the sedimentary layer. Choi et al. studied the influence of process parameters on the thickness of the deposited layer and reported that setting the powder mass

IOP Conf. Series: Materials Science and Engineering **894** (2020) 012012 doi:10.1088/1757-899X/894/1/012012

flow rate and the thickness per layer is critical to minimizing the difference between the preset layer height and the actual layer height [6]. In response, Do-Sik Shim et al. developed a process diagram for correlating the setting of the layer height and process parameters, and proposed a layer thickness setting method that was adapted to various processes [7]. This also revealed the influence of process parameters on material properties. Ruan et al. studied the empirical relationship between deposit height and laser scan rate under a constant laser power and powder feed rate [8]. However, as this empirical model only works under specific process conditions, the predicted deposit height results are not accurate when the experimental conditions are changed. Therefore, Ehsan Toysr et al. proposed a method of installing an optical charge-coupled device (CCD) sensor and performing feedback control in real time according to various experimental conditions by controlling the laser output using a pattern recognition algorithm [9]. However, as described in Do-Sik Shim et al., this method changes parameters that affect material properties such as the laser power and powder mass flow rate, so that the method may not yield an optimally shaped object [7].

Therefore, this study aims to not only improve the dimensional accuracy of the 3D metal lamination molding machine but also control the lamination height in real time without changing process parameters that affect material characteristics. The distance between the tip of the laser head and the molten pool (also called the standoff distance) causes an error with the input ideal distance as the number of layers increases. In this study, the height of the laser head nozzle is controlled in real time as the number of layers increases so that the height of the formed object is controlled. In order to observe the molten pool in real time, it is necessary to install a small camera on the head, so a jig for camera installation was constructed. We photographed the molten pool with the camera, and using the results of analyzing the captured image, the model was formed while correcting the head nozzle height as the number of layers increased. Finally, in order to confirm the usefulness of our modifications, we compared a product produced with this method to one produced by the conventional stacking method.

2. Control Method of Head Nozzle

2.1. Modeling Method by DED

In the DED method shown in Figure. 1, laser irradiation from the head nozzle is used to melt and stack the powder. A portion of the object to be shaped is irradiated with a laser to form a molten pool, toward which a metal powder is discharged together with a carrier gas (Ar) from a nozzle coaxial with the laser, and the head nozzle is moved in the laminating direction to perform lamination. The head nozzle also supplies shielding gas (Ar) that prevents air from entering into molten pool.

In order to increase the number of stacks, the head nozzle height is input to the 3D printer in accordance with the stacked height for each one-layer stack in the DED method. In conventional DED, the head nozzle height adjustment has been a constant value for each lamination cycle. As the number of lamination cycles increases, the difference between the height of the actually formed laminate and the ideal height calculated by the target parameters becomes larger. In other words, there is an error when comparing the dimensions of the finished object with the target dimensions.

2.2. Method for Measuring the Distance Between the Head Nozzle and the Molten Pool (Standoff Distance: SOD)

In this study, a camera is mounted on the head nozzle, and the distance between the head and the molten pool (SOD) is measured by triangulation from the image taken by the camera. Figure. 2 shows the positional relationship between the camera mounted on the head nozzle and the molten pool. From the laser irradiation angle and the camera position, the displacement x from the reference position of the molten pool can be calculated from Eq. 1.

$$x = \frac{dD}{f \sin\theta + d \cos\theta}. \tag{1}$$

From Figure. 2, the distance between the head nozzle and the centerline of the camera is A [mm], the distance between the center of the camera and the top of the molten pool is D [mm], and the angle

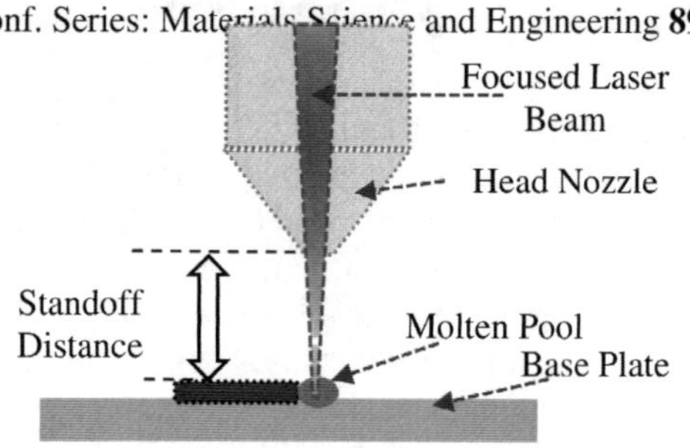

Figure. 1 Schematic of the DED stacking principle

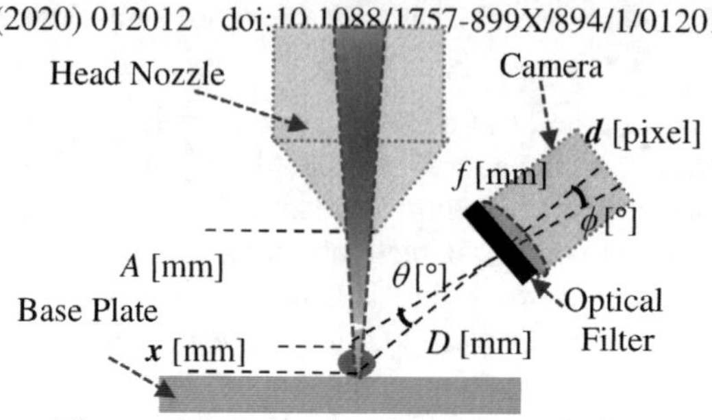

Figure. 2 Schematic of the triangulation

between the head nozzle and the center of the camera is , the camera focal length is f [mm], and the molten pool displacement from the camera is d [mm]. A, D, θ, and f are all fixed values. From Eq. 1, it can be seen that SOD is a curve for the shift of the molten pool in the camera head. Although the displacement x can be calculated from Eq. 1, when the displacement x is calculated from the position d of the molten pool in the camera image, the accuracy is reduced due to the camera position error and the irradiation angle error of the laser. Therefore, in this study, the relationship between displacement x and position d was acquired as calibration data, and displacement x was determined from position d.

Next, the method of acquiring calibration data is explained. First, we input the image (Figure. 3) taken by the camera into Python. Python can be used to transform the coordinate system with x [pixel] in the image and y [pixel] in the vertical and extract the coordinate system constructed by the image name and the RGB value of each coordinate. Furthermore, only specific RGB value ranges are extracted from the linked coordinate values. In this study, only the coordinate values for (R, G, B) = (255, 255, 255) are extracted. To obtain the center of the molten pool from the extracted coordinate values, it is assumed that the molten pool is a circle, with the y-axis being constant [10]. Finally, the location where the range of x is the largest is identified, and this midpoint is designated as the molten pool center. A calibration curve and standoff distance (SOD) can be obtained by plotting the molten pool center coordinates. The calibration curve can be approximated as a straight line because the value of d is sufficiently small.

The resolution p [mm/pixel] of the correction value x [mm] input to the laser layered metallurgical modeling machine can be obtained from Eq. 2 whenever the 1 [pixel] molten pool coordinate value changes.

$$p = \frac{x}{d}, \tag{2}$$

Where d [pixel] now is the change of the molten pool coordinate value in the image when the displacement of the molten pool changes x [mm].

2.3. How to Determine SOD

The displacement x from the reference position of the molten pool can be calculated by the above method. Since the optimal value of displacement x is 0, correction control is performed on the Z value (the value to be input to the 3D printer) of the formation path so that the displacement x becomes 0 when forming the layer above it.

3. Experimental Procedure

3.1. DED Processing Conditions

In this study, a shaped object of Inconel 625, a nickel-based heat-resistant alloy was fabricated using a laser metal forming and milling processing machine (LaserTec 65 3D, DMG Mori Co., Ltd.) as shown in Figure. 4. The experimental conditions such as SOD, laser power, and scanning speed were input to the multitasking machine by the NC program. The stacking conditions other than SOD are not changed. Regarding the measurement of the object, the shape of the object was reproduced using a Nikon P3D NC-2323S measuring system. Table 1 shows the experimental conditions. The VICOCO camera mounted on the head nozzle has a size of is 24 mm long, 26 mm wide, and 20 mm deep. The maximum resolution of the camera is 2.0×10^6 [pixel].

3.2. Acquisition of Calibration Date

The combined processing machine used in this study has an SOD reference value of 11 mm when performing the first layer of lamination, and it increases by 0.4 [mm] for every layer following. The calibration data were acquired by changing the SOD of the first layer from 9 [mm] to 13 [mm]. The relationship between the molten pool position *d* of the image taken

Table 1 Experimental conditions

Deposition Condition	Value
Laser power [W]	2,000
Feed rate [mm/min]	500
Powder flow [g/min]	14
Carrier Gas [l/min]	6
Shield Gas [l/min]	5
Powder material	Inconel 625
Powder size [μm]	53~150
Base plate	S45C

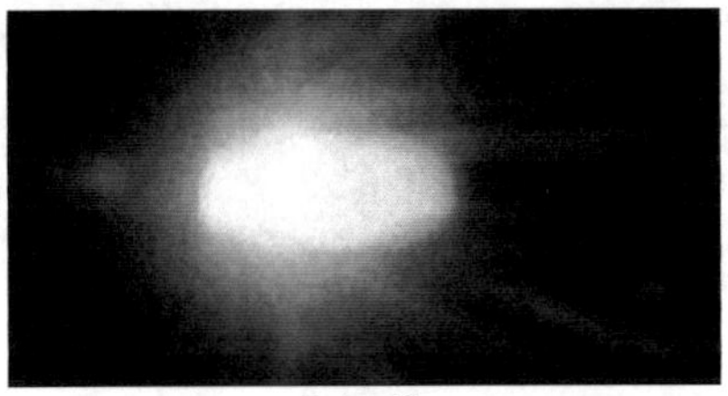

Figure. 3 Molten pool image

Figure. 4 LASERTEC 65 3D

with the camera and the SOD is as shown in Table 2, and the resolution in this study was determined to be $p = 0.049$ [mm / pixel].

3.3. Experiments Using the Correction Method

In experiments using the correction method, it is assumed that the position *d* of the molten pool is derived in real time; based on that, the displacement *x* of the head nozzle is also corrected and controlled. However, in this experiment, calculation of the position *d* and correction control of the displacement *x* could not be performed in real time because of the equipment, so after four layers were formed, the calculation of the position *d* and correction control of the displacement *x* was done manually, and the operation was repeated to form a single line of 25 [mm] up to 40 layers. In addition, in order to confirm the usefulness of this method, 40 layers of 25 [mm] in a single line were formed without correction control of displacement *x* (e.g., every time one layer is stacked, raise the nozzle by 4 [mm]).

3.4. Verification of the Experimental Results and the Usefulness of the Proposed Method

Figure. 5 shows the results of additive manufacturing with and without correction control of displacement *x*. The measurements of the cross section of the resulting shaped object are shown in Figure. 6. The model with correction had a height of 28.84 [mm], and the model without correction had a height of 20.26 [mm]. From this, it was confirmed that the correction control improves the shaping efficiency by 40%.

Figure. 7 shows the difference from the optimum value (11 mm) of the distance between the head nozzle and the molten pool with and without correction control. As shown in the figure, it can be seen that the correction control is optimally set at the corrected stage. That is, in the proposed method, the Z coordinate value of the head nozzle is known, and the distance from the head nozzle to the molten pool is maintained at 11 mm. Therefore, it is possible to perform molding knowing the height of the object. In this study, the moving distance of the head without correction was 16 [mm] (optimal modeling height) because 40 layers were laminated with 0.4 [mm] per layer, and the error rate with the actual object was 21%. The movement distance of the head with correction was 28.67 [mm] (optimal modeling height) as a result of changing 4 of the 40 layers, and the error rate with the actual object was 0.49%.

These results demonstrate that the proposed method can accurately mold an object with improved molding efficiency compared to conventional DED. Also, as can be seen from Figures. 5 and 6, the

Table 2 Molten pool coordinate values for head nozzle height change

SOD (Standoff Distance) [mm]	9	10	11	12	13
Molten pool coordinate value [mm]	967	982	989	1001	1013

Figure. 5 Modeling results

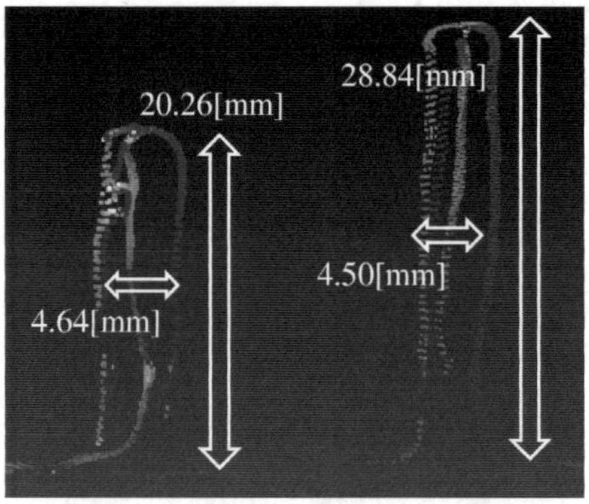

Figure. 6 Measurement results

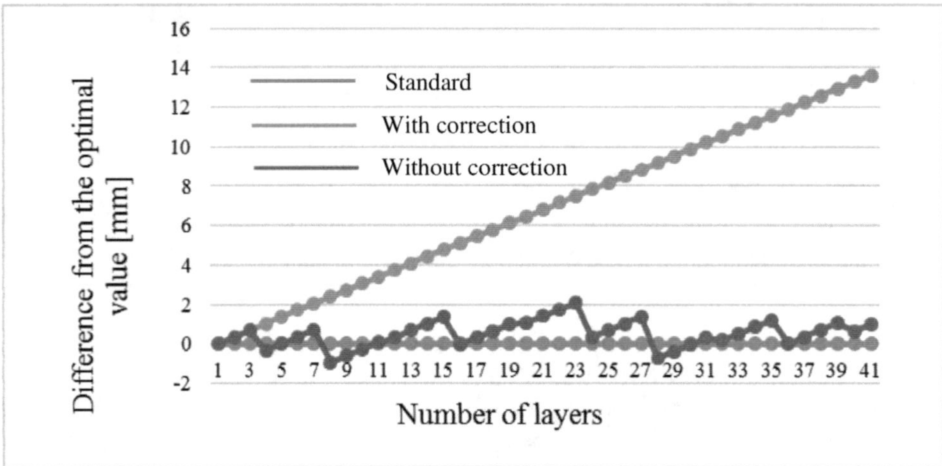

Figure. 7 Difference between SOD and optimum distance

width of the shaped object with correction was constant compared to the width of the shaped object without correction. This is because SOD was kept constant. Therefore, the proposed method's usefulness is confirmed (Figure. 7).

4. Conclusion

In this study, we propose a method of stable formation in DED and a method to keep SOD constant by monitoring the molten pool with a camera mounted on the head nozzle, and we compare the results of both methods experimentally. The proposed method is useful and forms an accurately shaped object with improved molding efficiency.

In future studies, we will improve the accuracy of the camera, fully automate the modeling method by inputting automatically corrected Z values for each layer, add the sensor function for measuring the molten pool temperature to the camera, and stably model the modeled object.

5. References

[1] H Kyogoku 2015 The Objective of a National Project of "Manufacturing Innovation Through Development of Next-Generation 3D Printers" *Journal of The Society of Instrument and Control Engineers* 54 5 pp 386

[2] X Wang M Zhang Z ZOU S Qu 2003 Microstructure and properties of laser –clad TiCp/Ni-based alloys composite coating *Chinese Journal of Mechanical Engineering* 39 pp 37-41

[3] J Sampedro I perez B Carcel J A Ramos V Amigo 2011 "Laser cladding of TiC for better titanium component *Physics Procedea* 12 pp 313 322

[4] S Zhou X Dai H Zheng 2012 Characteristics of structure and properties of WC-Ni60A coatings by laser cladding and laser-induction hybrid cladding *Chinese Journal of Mechanical Engineering* 48 pp 114 118

[5] Y Shi H Zhang S Liu C Xu 2003 Improving resistance of cobalt-based alloy coating formed laser fine cladding on high-hardness and microporous small parts *Chinese Journal of Mechanical Engineering* 39 pp 791 800

[6] J Choi Y Chang 2005 Characteristics of laser aided direct metal/material deposition process for tool steel *International Journal of Machine Tools and Manufasture* 45 pp 597 607

[7] Do-Sik Shim Gyeong-Yun Baek Jin-Seon Seo Gwang-Yong Shin Kee-Poong Kim Ki-Yong Lee 2016 Effect of layer thickness setting on deposition characteristics in direct energy deposition (DED) process *Optics & Laser Technology* 86 pp 69 78

[8] L Tang J Ruan R.G. Landers F Liou 2010 Direct three-dimensional layer metal deposition *Journal of Manufacturing Science and Engineering* 132 pp 064592_1 064592_6

[9] Ehsan Toyserkani, Amir Khajepour A mechatronics approach to laser powder deposition process 2006 *Mechatronics* 16 pp 631 641

[10] Masahiro Ueda David Carter 2018 A Study on Deposition Process Stability of DED-Type Additive manufacturing by Means of Monitoring Melt Pool Formation *The Proceeding of iAM-CNC 2018 Annual Meeting* pp 51 65

IOP Conf. Series: Materials Science and Engineering **894** (2020) 012013 doi:10.1088/1757-899X/894/1/012013

Thermal Performance of a Cross-flow Rotating Spray Bed

Moumita Sharma[a], Nilutpal Goswami[b] Avijit Bhowal[c]

Department of Chemical Engineering, Jadavpur University, Kolkata-700032, West Bengal, India

[a]moumitasharma21@gmail.com, [b]nilutpalgoswami13@gmail.com, [c]avijit.bhowal05@gmail.com

Abstract. This work examined the thermal performance of evaporative cooling of water by an air stream in an equipment wherein the liquid phase is sprayed between two coaxial disks. The contact between the two phases was in a cross-current mode. The performance of the spray bed in terms of thermal efficiency and evaporation rate of liquid were calculated by varying operational parameters like air flow rate, water flow rate, and rotational speed. Both the thermal efficiency (minimum = 0.183, maximum = 0.34) and evaporation rate (minimum = 0.009 kg/min, maximum = 0.032 kg/min) increased with rotational speed and air flow rate. However, the thermal efficiency decreased with liquid flow rate.

1. Introduction

Massive amount of waste heat is generated in air conditioning, power generation units, petrochemical, chemical processing industries etc that must be released continuously for proper functioning of the units [1]. A proper coolant is required to reject this heat from the system. Water is considered as a suitable coolant due to its high heat capacity, easy availability, minimal cost and non degrading nature. Mass transfer devices like cooling towers are used to reuse the water. Cooling in this equipment is based on the principle of Evaporative cooling [2]. The process involves introduction of warm water at the top of the tower. In these towers, the air and the warm water are contacted directly. The liquid comes down under the action of terrestrial gravity through fills or packings. Transfer of sensible heat as well as the latent heat takes place between the two phases, the later because of the evaporation of water. The flow may be countercurrent or crosscurrent.

The thermal characteristics of countercurrent cooling tower and the impact of different filling materials had been studied by various researchers [3-6]. Gonzalez et al. [7] has optimized a mechanical draft countercurrent cooling tower for minimizing the cost. Grobbelaar et al. [8] studied the performance characteristics of a cross and countercurrent wet cooling tower using trickle fills. Spray cooling technique has been successfully utilized in various industries like Gas Turbine fogging, food refrigeration etc [9]. Sun et al. [9] has studied the effect of spray cooling in a naturally draft drying cooling tower. Tissot et al. [10] has performed an experimental study on air cooling by spray in a heat exchanger. Si et al. [11] has studies the effect of cross-flow on the structure of the spray, droplet diameter and velocity distributions in a gasoline engine.

In this study, gravitational force used in traditional cooling towers to drive the liquid is replaced by centrifugal force which is few hundred times to that of gravity. There is no previous literature to indicate the efficacy of this process. Therefore the objective of this work was to evaluate the thermal performance and cooling effectiveness in a cross current spray bed.

2. Materials and Methods

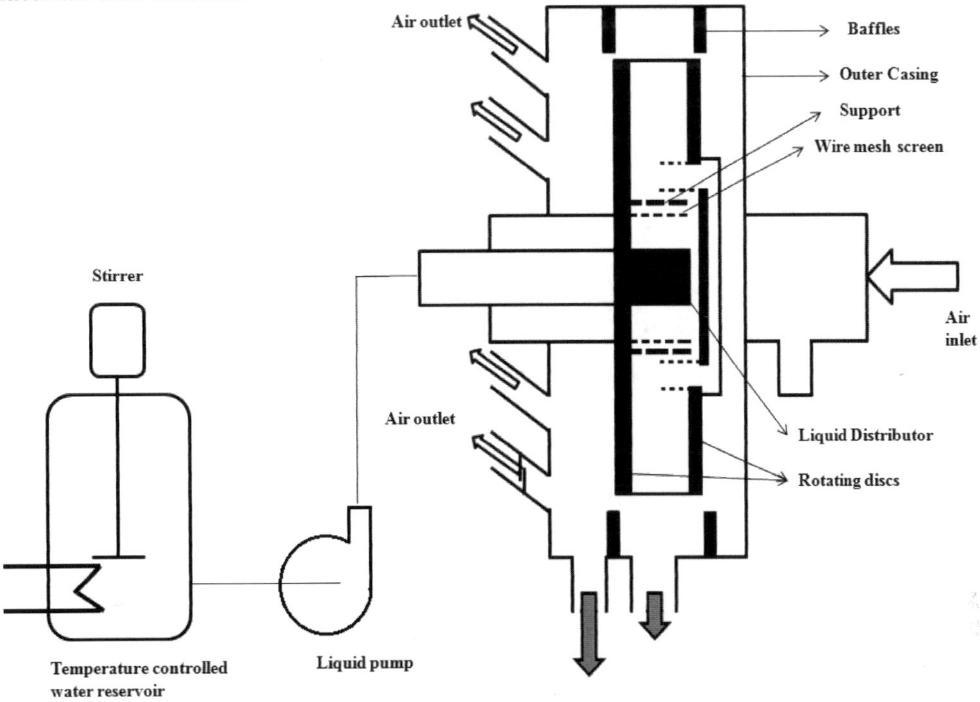

Figure.1. Experimental Set Up

The experimental set up is provided in a schematic form in figure 1. Warm water was contacted with the air inside the rotor under the action of centrifugal force. Two thin strips of metals were welded onto thin circular strips in both sides of the rotor. The rotor was connected to an AC motor through a shaft that rotated along the horizontal axis. The insulated casing has a diameter of 238mm and an axial length of 75mm. In order to minimize the bypassing of air within the rotor, circular baffles were attached to either sides of the casing wall.

Air was introduced inside the rotor by a blower. Water was heated to 500C in a reservoir provided with heating equipment and a stirrer. It was pumped into the rotor through a stationary distributor (provided with 24 holes of 1mm each). The liquid get dispersed in form of jets from the distributor and strikes the wire mesh strip aligned perpendicular to the direction of flow. The liquid was thrown in the outward radial direction inside the rotor due to strong centrifugal force and was discharged at the bottom through an opening on the casing wall. The temperature of the water and air entering as well as leaving the contactor were measured using a thermocouple connected to a digital display. The humidity of the inlet and outlet air was measured using a digital humiditymeter. Baffles were provided in each of the four air outlets to prevent the carryover of water.

3. Results and Discussions

Simultaneous heat and mass transfer performance of the contactor was estimated in terms of thermal efficiency (η) and rate of Evaporation ($\dot{e}$). The measurement of thermal efficiency inside the contactor is illustrated in Figures 2 and 3. Thermal efficiency (η) is defined as

$$\eta = \frac{t_{w,i}-t_{w,o}}{t_{w,i}-t_{wb,i}} \tag{1}$$

where, $t_{w,i}$, $t_{w,o}$ represents the inlet and outlet temperature of water respectively, and $t_{wb,i}$ is the wet bulb temperature of inlet air. It is the ratio of the experimentally determined cooling range, Δt_c $\left(=t_{w,i}-t_{w,o}\right)$ and the cooling limit.

IOP Conf. Series: Materials Science and Engineering **894** (2020) 012013 doi:10.1088/1757-899X/894/1/012013

The effect of thermal efficiency with respect to rotational speed at different liquid flow rates and constant gas flow rate of 0.03 kg/s is shown in figure 2. It was observed as the rotational speed was increased from 300 to 1100 rpm, there is an increasing trend in the thermal efficiency. This is due to fact that high rotational speed results in more shearing force that reduces the size of the droplets and enhances the gas-liquid interfacial area to a great extent [12]. The maximum thermal efficiency obtained was 0.31. With increase in the liquid flow rate from 0.0125 kg/s to 0.029 kg/s, there is a gradual decrease in the efficiency.

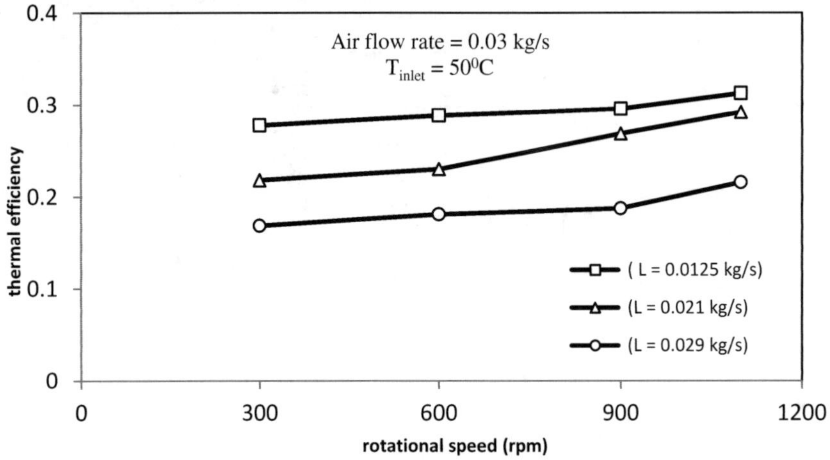

Figure.2. Effect of rotational speed on thermal efficiency

The influence of liquid to gas mass flow rate ratio and rotational speed on thermal efficiency in the spray bed contactor is illustrated in Figure 3 at constant liquid flow rate of 0.0125 kg/s. The thermal efficiency is noted to increase with rotational speed for all the operating conditions. η increased from 0.28 to 0.34 at $\frac{\dot{L}}{\dot{G}}$ ratio of 0.31 in the spray bed as the rotational speed was increased from 300 to 900 ($\dot{L}$= 0.0125 kg/s).

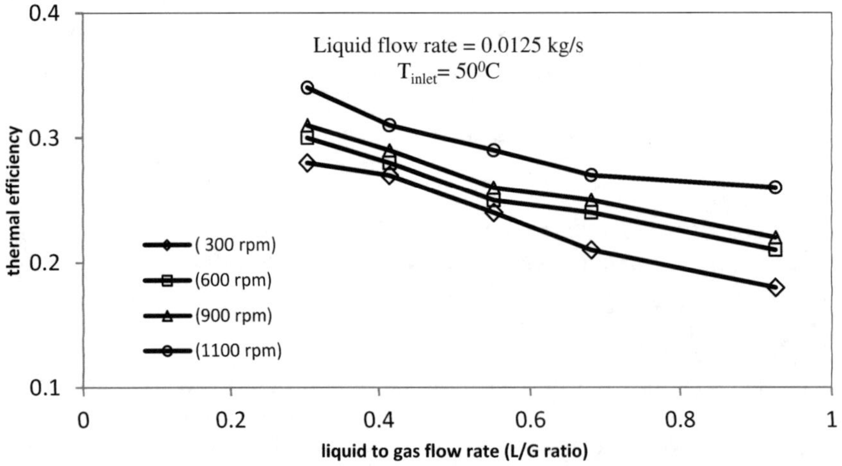

Figure.3. Effect of liquid to gas flow rate on thermal efficiency

The effect of air flow rate on the thermal efficiency (shown in figure 4) was studied at constant flow rate of liquid (0.0125 kg/s) for different rotational speed. With the increase in the air flow rate from 0.0128 kg/s to 0.03 kg/s there was an increase in the efficiency from 0.18 to 0.31.

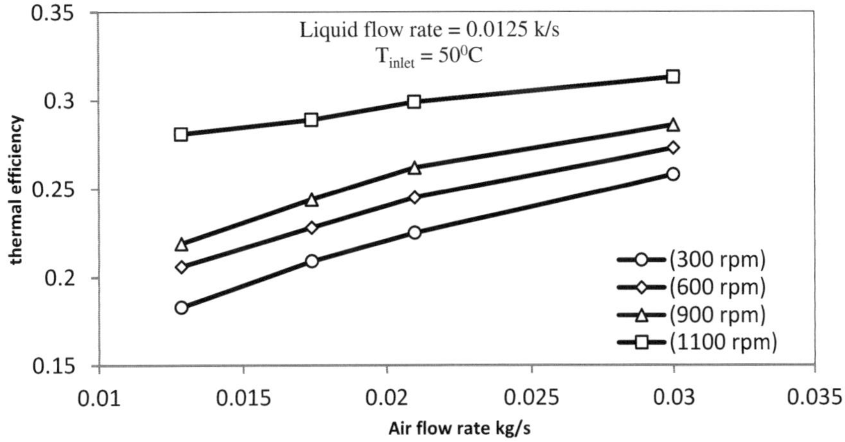

Figure.4. Effect of air flow rate on thermal efficiency

The evaporation rate of water ($\dot{e}$) was estimated from the difference in moisture content between the inlet and outlet air, and the air flow rate.

$$\dot{e} = \dot{G}(Y_{out} - Y_{in}) \qquad (2)$$

The variation of rotational speed with the evaporation rate is shown in figure 5. The evaporation rate increased with the rotor speed and liquid flow rate.

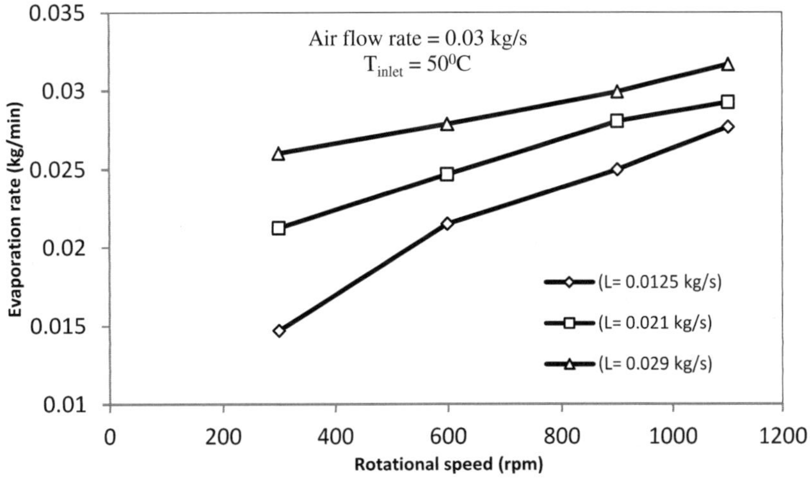

Figure.5. Effect of rotational speed on evaporation rate

As the liquid flow rate was increased from 0.0125 to 0.029 kg/s with constant air flow rate 0.03 kg/s, there was an increase in the evaporation rate from 0.009 kg/min to 0.032 kg/min. Figure 6. depicts the effect of air flow rate on the evaporation rate at different rotational speed at constant liquid flow rate 0.0125 kg/s. With increase in the air flow rate from 0.0128 kg/s to 0.03 kg/s, increase in evaporation rate was observed from 0.009 kg/min to 0.028 kg/min.

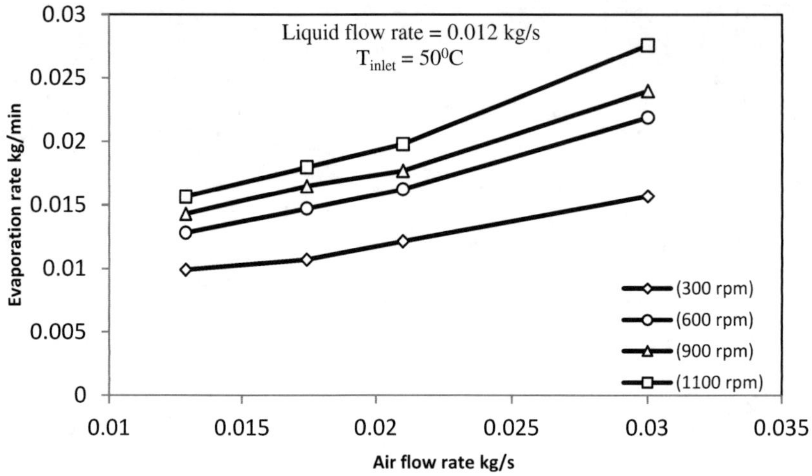

Figure.6. Effect of air flow rate on evaporation rate

Smaller water droplets or thinner ligaments are formed on increase of rotational speed [13]. This increases the gas-liquid surface area. Increase in liquid flow rate leads to better distribution of liquid over the rotor from the liquid distributor opening as more of the distributor openings become effective. This increase the surface area for evaporation of water and sensible heat transfer which results in high evaporation rate. The increase in air flow rate facilitates evaporative cooling by increasing sensible heat uptake capacity of the air particles in contact with the warm water.

The temperature change of water on account of only evaporation (Δt_e) can be calculated by $\frac{\dot{e}\lambda}{C_w \dot{L}}$ where λ is the latent heat of vaporization of water. The contribution of Δt_e to the experimentally determined cooling range of water varied between 75% and 90% in these contactors. Thus, cooling of water is achieved primarily by latent heat supplied by liquid for evaporation.

4. Conclusion

The thermal performance of a rotating spray bed was evaluated in a cross current mode for observing the evaporative cooling in water. Cooling is primarily obtained by the evaporation of water. The efficiency and evaporation rate for the range of condition studied varied between 0.18 and 0.34, and 0.009 kg/min and 0.032 kg/min respectively.

5. Nomenclature

$\dot{e}$	evaporation rate of water (kg/min)
$\dot{G}$	gas mass flow rate (kg/s)
$\dot{L}$	liquid mass flow rate (kg/s)
C_w	heat capacity of water (KJ/kg °C)
t_w	temperature of water (°C)
t_{wb}	wet bulb temperature of air (°C)
Subscript	
i	inlet
o	outlet
Greek letters	
λ	latent heat of vaporization of water (KJ/kg)
η	thermal efficiency

Δt_e	temperature change in water due to evaporation (°C)
Δt_c	cooling range of water (°C)

Acknowledgement
The financial assistance provided under the University Grants Commission (India), UPE-Phase II scheme of Jadavpur University is cordially appreciated.

6. References
[1] M. Lemouari, M. Boumaza, A. Kaabi: Energy Convers. Manag. Vol. 50 (2009), p. 1610-1617.
[2] F. Gharagheizi, R. Hayati, S. Fatemi: Energy Convers. Manag. Vol. 48 (2007), p. 277-280.
[3] J-U-R. Khan, M. Yaqub, S.M. Zubair: Energy Convers. Manag. Vol. 44 (2003), p. 2073-2091.
[4] E. Hajidavalloo, R. Shakeri, M.A. Mehrabian: Energy Convers. Manag. Vol. 51 (2010), p. 1298-1303.
[5] T. Muangnoi, W. Asvapoositkul, P. Hungspreugs: Appl. Thermal Eng. Vol. 69 (2014), p. 165-176.
[6] S.V. Bedekar, P. Nithiarasu, K.N. Seetharamu: Energy Vol. 23 (1998), p. 943-947.
[7] M. S-Gonzalez, J. M. P-Ortega, A. J-Gutierrez: Chem Eng. Res. Des. Vol. 88 (2010), p. 614-625.
[8] P.J. Grobbelaar, H.C.R. Reuter, T.P. Bertrand: Appl. Thermal Eng. Vol. (2013), p. 475-484.
[9] Y. Sun, Z. Guan, H. Gurgenci, K. Hooman, X. Li, L. Xia: Int. J. Heat Mass Transf. Vol. 110 (2017), 113-131.
[10] J. Tissot, P. Boulet, A. Labergue, G. Castanet, F. Trinquet, L. Fournaison: Int. J. Therm. Sci. Vol. 60 (2012), 23-31.
[11] Z. Si, Y. Ashida, N. Shimasaki, K. Nishida, Y. Ogata: Fuel Vol. 234 (2018), 592-603.
[12] L. Patridge, D.C.Y. Wongs, M.J.H. Simmons, E.I. Parau, S.P. Decent: Chem. Eng. Res. Des. Vol. 83 (2005), p. 1267-1275.
[13] D.C.Y. Wongs, M.J.H. Simmons, S.P. Decent, E.I. Parau: Int. J. Multiphase Flow Vol. 30 (2004), p. 499-520.

IOP Conf. Series: Materials Science and Engineering **894** (2020) 012014 doi:10.1088/1757-899X/894/1/012014

Differences in Polysaccharide Containment Obtained by Various Pulsed Light Irradiated Waste Mushroom Base of Yellow *Flammulina Velutipes*

Shu-Yao Tsai[1], Pei-Han Wu[1], Chun-Ping Lin[1,2,]*

[1] Department of Food Nutrition and Health Biotechnology, Asia University, 500, Lioufeng Rd., Wufeng, Taichung, Taiwan 41354, ROC
[2] Department of Medical Research, China Medical University Hospital, China Medical University, 91, Hsueh-Shih Rd., Taichung, Taiwan 40402, ROC

*E-mail: cp.lin@asia.edu.tw; chunping927@gmail.com (C.-P. Lin)

Abstract. The goal of this study was applied on the pulsed light of physical method, to promote the polysaccharide containment of mushroom base of Yellow *Flammulina velutipes* (Jinhua mushroom). Jinhua mushroom is a specific edible and medicinal mushroom in Taiwan that contains a variety of nutrients and bioactive components such as vitamin, carbohydrate, dietary fiber, protein, ergosterol, amino acid and umami constituents acknowledged to be beneficial for human health. Jinhua mushroom, whose appearance is also similar to white *Flammulina velutipes* but the functional ingredients are more than it. The polysaccharide of Jinhua mushroom base, which was irradiated the 0–80 pulses of pulsed light. Interestingly, the polysaccharide molecular weight was in the range of 3.56 to 4.30 x 10^6 Da of base with pulsed light (0–80 pulses), along with increasing pulsed light irradiated pulses, the polysaccharides' viscosity (cPs) was increased. Meanwhile, we determined the moisture adsorption ability and thermal decomposition of the Jinhua mushroom base's polysaccharide, which could be applied to a design during cosmetic materials and biomedicals. Overall, this study is a complete, forward-looking and innovative research project in agricultural waste recycling.

1. Introduction

Yellow *Flammulina velutipes* (also call Jinhua mushroom) has become more popular to Taiwanese in recent years; its distinct texture and unique flavor are utterly different from white *Flammulina velutipes*. According to the literature, the biological activities are associated with the molecular weight distribution of polysaccharide. In addition, *Pholiota nameko* polysaccharide was purified by fractional precipitation, already demonstrated to have the antioxidant activity of polysaccharides [1].

This study was aim of the polysaccharides of Jinhua mushroom base by pulsed light irradiation of physical method. We applied physical treatment to irradiate the Jinhua mushroom base with pulsed light to study its polysaccharides content, to compare the base by different doses of pulsed light irradiated, and then contrasting the polysaccharide of ingredients, from which could be learned the reliable irradiation pulses of Jinhua mushroom base. In addition, the polysaccharides were obtained by hot water extracted of the base of Jinhua mushroom, which will irradiate at 0 (blank experimental control was unirradiated), 10, 30, 50, and 80 pulses by pulsed light [2].

Differential scanning calorimetry (DSC) was used to determine the energy changes of the moisture desorption and thermal decomposition of the polysaccharide of Jinhua mushroom base [3]. A novel approach to discover the differences in polysaccharides obtained from different doses irradiation of the Jinhua mushroom base, that includes the viscosity characteristics, moisture desorption ability, and thermal decomposition properties. Moreover, we also compared the differences in structural functional groups after irradiation by infrared spectroscopy analysis. Therefore, this study can be applied to design during food processing, cosmetics processing, heat treatment, and storage conditions.

2. Materials and methods

2.1. Materials

Yellow *Flammulina velutipes* (strain with yellow colour) were donated by Da-Tian Farm Co. Ltd. (Changhua, Taiwan). The base (YB0) of ca. 4.0 cm in length is the waste part (see Figure. 1.).

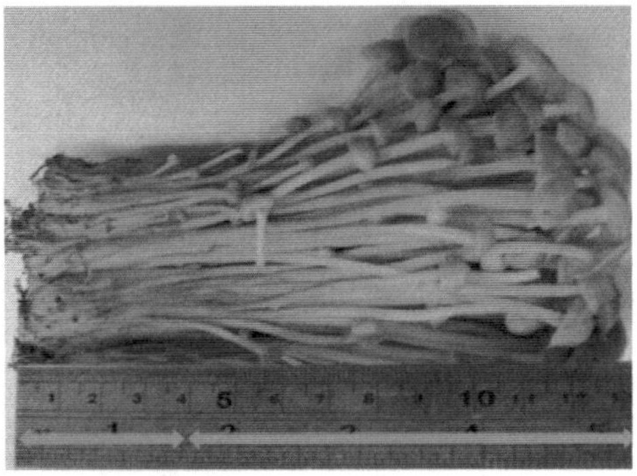

Figure 1. Photograph of sample treatment.

2.2. Irradiation

Samples were subjected to irradiation using a Xenon RC-801 pulsed light system (Xenon Corp.) employing the RC-847 cabinet, lamp housing LH-840 and C-type 16″ linear lamp (190–700 nm). The apparatus generated three pulses per second at 169 J pulse^{-1} [2]. An average dose was 11.50 kJ m^{-2} pulse^{-1} (Koyyalamudi et al. 2011). Each mushroom base was thinly scattered, placed in the middle of a stainless-steel plate and irradiated for 0, 10, 30, 50, and 80 pulses.

2.3. Jinhua mushroom base polysaccharides

The samples after irradiation were freeze-dried and ground into a fine powder (60 mesh) by using a mill (Retsch ultracentrifugal mill and sieving machine, Haan, Germany). Different doses of pulsed light irradiation of 0 (un-irradiated), 10, 30, 50, and 80 pulses, which are YB0, YB10, YB30, YB 50, and YB 80, respectively, were extracted with water in a 1: 30 (w/w) ratio at 121°C for 15 min. The mixture was cooled to room temperature and filtered by Whatman No. 4 filter paper. The combined filtrate was dialyzed by using a Cellu Sep T2 tubular membrane (MWCO: 6,000-8,000, Membrane Filtration Products, Inc., Seguin TX, USA) for 24 h. The retentate was concentrated to a small volume and then mixed with three volumes of 95 % ethanol [2,3]. The precipitate thus obtained was lyophilized and ground using a Retsch mill to obtain coarse powder of hot water extracted polysaccharides from 0, 10, 30, 50, and 80 pulse sample.

2.4. Jinhua mushroom base polysaccharides viscosity analysis

All samples were evaluated at room temperature, conducted by sine wave Vibro-viscometer A&D SV-10 (A&D Company, Limited, Tokyo, Japan), that was under constant frequency at 30 Hz and amplitude less than 1 mm [2].

2.5. Jinhua mushroom base polysaccharides molecular distribution analysis

The polysaccharide molecular weight distributions were applied on the high performance liquid chromatography (HPLC) with a Hitachi HPLC apparatus equipped with a 2490 refractive index (RI) detector. Polysep-GFC-P 4000 and Polysep-GFC-P 1000 (7.8 × 300 mm, Phenomenex) were employed after connection in series, and the temperature of the column was kept at 30 °C. The sample concentration was 1.0 mg mL^{-1}, and its injection volume was 10 μL. The eluent was ultrapure water, and the flow rate was 1.0 mL min^{-1} [2,3].

2.6. Jinhua mushroom base polysaccharides infrared spectroscopy analysis

All sample was added into ca. 150 mg of dried KBr; the sample was fully mixed with KBr after grinding, and then put in a tablet machine pressed into a disc. Infrared spectrum were acquired with a Bruker Alpha FTIR spectrometer. Collections of spectra were in the transmission mode with an unpolarized light beam, at a resolution of 4 cm^{-1} with eight scans, and a spectral range of 4000–400 cm^{-1} [4].

2.7. Jinhua mushroom base polysaccharides DSC tests

DSC TA Q20-RCS90 (TA Instruments, USA) for conducting DSC analysis, the samples were sealed in 20 μL aluminum pans, and then the lid was pressed onto the crucible using the pressure of a heavy mechanical force and the seal tightened the crucible [2,3] Approximately 2.0 mg of the sample was used for received the thermal analysis experimental data. Non-isothermal tests of the heating rate selected for the programmed temperature ramp were 2, 4, 6, and 8 °C min^{-1} for the range of temperature rise chosen from 30 to 300 °C for each thermal analysis.

3. Results and discussion

3.1. Polysaccharide viscosity and molecular distributions of different dose pulsed light irradiations

Pulsed light irradiation induced significant changes in the viscosity of the polysaccharide solution as shown in Table 1. As the irradiation dose increased from 0 to 80 pules, the YB0–YB80 polysaccharide solution showed a significant increase in viscosity. The viscosity of the YB0–YB80 polysaccharide was from 1.77 to 1.86 cP (mPa·s), whereas the viscosity of the high viscosity 1.86 cP was irradiated at 80 pulses. The increase in the viscosity is attributed to the increase in degree of polymerization of polysaccharide [2]. In addition, the irradiation dose increased from 0 to 80 pules, as the dose of 80 pulses, the polysaccharides had the greater molecular weight 4.30 x 10^6 Da.

Table 1. Yield, viscosity and molecular weight of crude polysaccharides of pulse light irradiated the base of yellow *F. velutipes*

Sample	Yield (%)	Viscosity (cPs)	Molecular weight (10^6Da)
YB0	5.47^a	1.77^d	3.56
YB10	4.49^e	1.80^c	3.93
YB30	4.63^d	1.83^b	2.95
YB50	4.75^c	1.86^a	3.84
YB80	5.14^b	1.86^a	4.30

*Means with the same letter within a column are not significantly different (P>0.05).

3.2. Polysaccharides of different dose pulsed light irradiations by infrared spectroscopy analysis

Figure 2 displays the hydroxyl (–OH) groups of the stretching-vibration absorption peak wavenumbers, which were 4000–3850 cm^{-1} and 3700–3100 cm^{-1}; of which the characteristic functional group of polysaccharide hydroxyl group clearly exhibited in the infrared spectrum. In addition, the polysaccharides of intermolecular hydrogen bonds 3650–3200 cm^{-1} were particularly noticeable. For the O–H stretching-vibration and bending-vibration absorption peak the wavenumbers were 3000 cm^{-1}, 1550 cm^{-1}, 1400 cm^{-1} and 980 cm^{-1}, respectively. The C–O stretching-vibration absorption peak wavenumber was 1650 cm^{-1}, and C–O–H bending-vibration absorption peak wavenumber was 1580 cm^{-1}, respectively, which is also clearly shown in the spectrogram. From comparison of Table 1 and Figure 2, the results of polysaccharide molecular weight distributions and infrared analysis were consistent. We obtained the judgment, as Jinhua mushroom was irradiated by the pulsed light reaction, that it initially increases the molecular weight of the polysaccharide, along with continuous irradiation producing an effective resistance of subsided phenomenon; then continued increasing the dose pulsed light irradiation kept increasing the molecular weight of the whole polysaccharide.

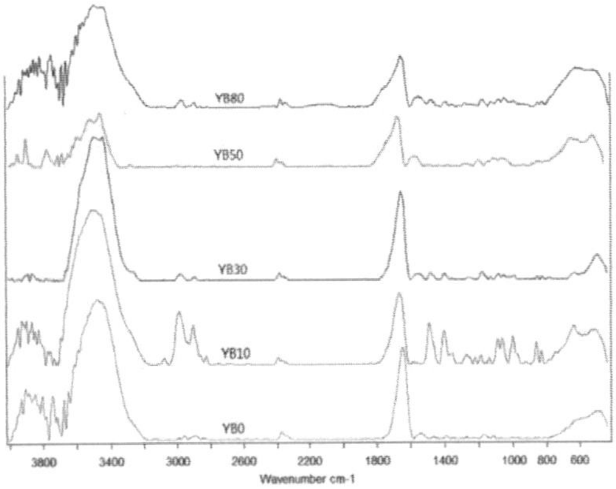

Figure 2. Different doses of pulsed light irradiations infrared spectrogram.

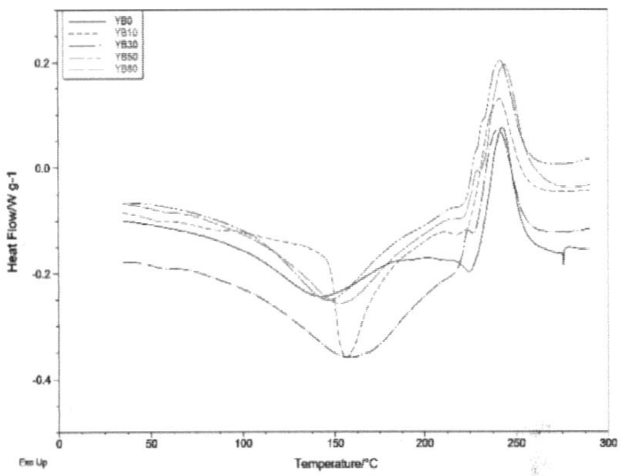

Figure 3. Selected DSC tests of the YB0–YB80 polysaccharides at heating rate 4 °C min^{-1} of the range of temperature rise chosen from 30–300 °C

Table 2. DSC analysis of YB0–YB80 polysaccharides with heating rate 2, 4, 6, and 8 °C min^{-1} of the range of temperature rise chosen from 30–300 °C.

Sample[a]	Heating rate[b]	desorT_o[c]	desorT_p[d]	desorΔH[e]	decompT_o[f]	decompT_p[g]	decompΔH_d[h]
YB0	2	81.80	127.54	238.20	227.44	237.70	72.69
	4	93.45	140.81	87.83	230.37	242.91	66.75
	6	91.13	130.29	247.70	240.80	251.77	118.20
	8	87.93	161.39	228.20	225.38	234.47	87.98
YB10	2	99.38	158.63	212.90	211.01	226.03	85.77
	4	149.20	156.56	125.30	226.36	241.12	75.27
	6	121.71	164.12	184.20	218.85	227.17	75.00
	8	91.62	138.15	214.80	243.23	254.47	102.1
YB30	2	91.93	133.89	214.80	223.78	233.95	71.01
	4	94.18	159.45	193.70	227.00	240.31	74.40
	6	104.48	151.31	283.60	232.39	245.69	101.70
	8	103.37	151.52	202.30	239.99	251.92	101.90
YB50	2	87.10	129.68	189.00	228.64	238.31	79.09
	4	95.66	146.37	182.80	227.08	241.42	79.66
	6	115.81	169.83	185.80	224.86	244.70	74.71
	8	98.87	147.18	175.10	237.06	252.82	115.20
YB80	2	84.19	131.38	226.10	228.40	239.80	98.66
	4	100.03	152.96	162.90	230.87	243.84	87.40
	6	105.44	146.67	181.50	238.46	250.72	117.60
	8	93.65	150.27	207.40	242.53	255.83	108.20

[a]Sample name. [b]Heating condition: non-isothermal (°C min^{-1}) DSC test. [c]Onset temperature (°C) of desorption reaction of DSC non-isothermal test. [d]Peak temperature (°C) of desorption reaction of DSC non-isothermal test. [e]Desorption reaction of enthalpy (kJ kg^{-1}) by DSC test. [f]Onset temperature (°C) of decomposition reaction of DSC non-isothermal test. [g]Peak temperature (°C) of decomposition reaction of DSC non-isothermal test. [h]Decomposition reaction of enthalpy (kJ kg^{-1}) by DSC test.
Remarks: Standard deviation: temperature accuracy: +/- 0.1; temperature precision: +/- 0.05; calorimetric reproducibility: +/- 1 mass%; sensitivity: 1.0 uW.

3.3. Polysaccharides of different dose pulsed light irradiations by DSC thermal analysis

From Table 2, the initial conversion (onset) temperatures of moisture desorption and thermal decomposition were ca. 93 and 230 °C (selected YB0 at heating rate 4 °C min^{-1}), respectively. The thermal decomposition behavior was by heat treatment, that also obtained the Jinhua mushroom base polysaccharide's peak maximum temperature of ca. 240 °C (see Figure 3). Furthermore, Table 2 shows

YB0–YB80 polysaccharides of Jinhua mushroom base have superior absorbent properties, as the YB0–YB80 desorption of moisture was heated more than 140 °C to reach equilibrium. It is essential for the polysaccharides of Jinhua mushroom involving the hydroxyl group (–OH) and the ether group (–O–) that they have good moisture absorbance and high sensitivity to ambient humidity [2]. Among all polysaccharide samples, the YB80 have the greater molecular weight 4.30 x 10^6 Da and viscosity 1.86 cP. The results of DSC thermal analysis proved again that polysaccharides of Jinhua mushroom base could increase the molecular weight by pulsed light irradiation. As also, the Jinhua mushroom base with different doses of pulsed light of 0, 10, 30, 50, and 80 pulses, along with increasing the temperature of moisture desorption to reach equilibrium, as YB80 illustrated its equilibrium temperature more than 152 °C. Obviously, after pulsed light irradiation, the polysaccharides' structure was changed in this study.

4. Conclusion

Comparisons of the different doses of pulsed light of 0–80 pulses showed that the Jinhua mushroom, acquired and demonstrated after pulsed light, the polysaccharides' structure was changed. Overall, the result of viscosity, molecular distribution, infrared spectroscopy, and DSC thermal analysis of different dose pulsed light irradiations was demonstrated; the Jinhua mushroom was irradiated by the pulsed light reaction. In the initial period it could be increased the molecular weight of the polysaccharide, during continuous irradiations producing an inhibition response, but continued increasing the dose pulsed light irradiation, which promoted the molecular weight for whole polysaccharide. This study is a forward-looking and innovative research project of the improved functional ingredients of Jinhua mushroom in food processing and heat treatment. In the future, our goal is to enhance the processing technology for health foods of edible and medicinal mushrooms to find improved processing technology and energy efficiency, a less polluting method and improved functional ingredients technology in the food, medicines and food safety areas.

Acknowledgments

Authors are indebted to the Ministry of Science and Technology (MOST), Taiwan, ROC under the contract No.: MOST 107-2221-E-468-007-MY2 and MOST 105-2221-E-468-001-. In addition, we also thank to SD Biotech Inc. in Taiwan for technical support on the pulsed light experiments and Da-Tian Farm Co. Ltd. (Changhua, Taiwan) donated the Jinhua mushroom.

5. References

[1] Hu YN, Sung TJ, Chou CH, Liu KL, Hsieh LP and Hsieh CW 2019 *Antioxidants.* **8** 298
[2] Tsai SY, Hwang BF, Wang YH and Lin CP 2017 *J. Therm. Anal. Calorim.* **127** 469
[3] Lin HY, Tsai SY, Tseng YL and Lin CP 2015 *J. Therm. Anal. Calorim.* **120** 439
[4] Tsai SY, Lin HY, Hong WP and Lin CP 2017 *J. Therm. Anal. Calorim.* **130(3)** 1357

IOP Conf. Series: Materials Science and Engineering **894** (2020) 012015 doi:10.1088/1757-899X/894/1/012015

Failure of Standard Duplex Stainless Steel X2CrNiMoN22-5-3 under Corrosion Fatigue in Geothermal Environment

Marcus Wolf[1,2] and Anja Pfennig[1,*]

[1] Dept. Mechanical Engineering, Applied University of Berlin HTW, Berlin, Germany
[2] Federal Institute of Materials Research and Testing BAM, Berlin, Germany

E-mail: anja.pfennig@htw-berlin.de

Abstract. Corrosion fatigue is well known to lower the lifetime expectancy of high alloyed steels. Duplex stainless steel X2CrNiMoN22-5-3 is a promising candidate to withstand both, corrosion and mechanical stress. Therefore, the corrosion fatigue (CF) was investigated in the Northern German Basin electrolyte at 369 K using a specifically designed corrosion chamber. CF damage was clearly related to lateral grain attack within corrosion pit cavities located perpendicular to the load applied. Additionally, multiple fatigue cracks and preferable deterioration of austenitic phase and intact ferritic phase. A delta-like micro crack structure and a curved path characterizes crack termination with little to no base metal deterioration. Crack initiation may be due to early pit formation resulting in depassivation but also due to local depassivation then resulting in pit formation – both initiation mechanisms lead to crack propagation and early failure.

1. Introduction

Because materials for future geothermal energy extraction applications in the energy sector such as delivery pumps as well as components for conveying and filtration have to meet new requirements due to being loaded mechanically and at the same time exposed to highly corrosive environments. Frequency of cyclic loading, temperature and chloride concentration of the corrosive media influence corrosion processes when mechanical stress is simultaneously applied [1]. Degradation by corrosion fatigue mechanisms increases for steels with low chromium content [2] and in the presence of chlorides [3], hydrogen sulphide (H_2S) [4] as well as CO_2 [5]. For high alloyed steels increasing temperature and mechanical load as well as decreasing pH enhance corrosion fatigue phenomena [6–9] whereas chromium content and internal compressive surface stress increases the endurance limit.
Standard duplex stainless steel DSS X2CrNiMoN22-5-3 (AISI 2205) used in heat exchangers, refinery towers, desalination plants and other operational units of chemical and petroleum industries [10] reveals high mechanical strength [10] resistance to stress corrosion cracking [11] and corrosive mediums [12,13].

Corrosion fatigue was tested in the geothermal brine of the Northern German Basin at 98 °C from S_a 175 to 325 MPa [14-18]. The fatigue limit of duplex stainless steel X2CrNiMoN22-5-3 in non-corrosive environment of is about 485 MPa (P_f =50%, push/pull) after a test duration of 10^7 cycles [15-18]. Under corrosion fatigue at free corrosion potential X2CrNiMoN22-5-3 reached the maximum number of cycles of (9.2 x 10^6) at a stress amplitude of 240 MPa (Figure 1, left) [15-18]. Main surface

and pit precipitation phases are mainly carbonates or hydroxides ($FeCO_3$, FeOOH) independent of the original surface roughness [7,16].
Due to the rather low scatter ranges (technical surface: TN=1:1.35, polished surface: TN=1.1.95, Figure 1 right [15-18]) it may be assumed that the failure mechanism remains the same regardless of surface condition. This paper reveals microstructural analysis to provide deeper understanding of crack initiation, propagation and final failure of X2CrNiMoN22-5-3 under conditions similar to those occurring in geothermal energy production.

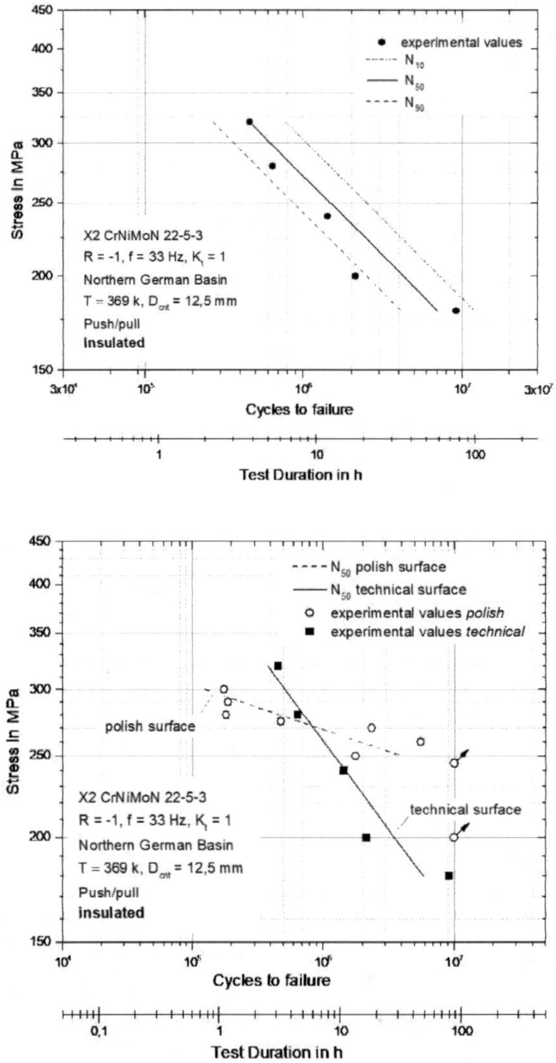

Figure 1. S-N-curves of X2CrNiMoN22-5-3. Left: using an insulated experimental setup [15-18], right: comparison of corrosion fatigue behaviour as a function of surface structures [15-18]

2. Materials and Methods

The corrosion fatigue tests have been performed using duplex stainless steel X2CrNiMoN22-5-3 (Table 1), manufactured by Deutsche Edelstahlwerke GmbH (yield strength: 672 MPa, tensile strength: 854 MPa). The fabrication was accomplished by means of continuous casting with a skiving finish where the alloy subsequently underwent tempering at 1050 °C, followed by water quenching ensuring the ferrite/ austenite balance and the absence of intermetallic phases [16].

Table 1. Chemical composition of the specimens tested

phases	C	Si	Mn	P	S	Cr	Mo	Ni	N
α & γ** 0,023	0,48	1,83	0,024	0,008	22,53	2,92	5,64	0,146	
α* 0,02	0,55	1,59	-	-	24,31	3,62	3,81	0,07	
γ* 0,03	0,47	1,99	-	-	20,69	2,17	6,54	0,28	

[a] spark emission spectrometry SPEKTROLAB M (Spektro)
[b] Electron Probe Microanalyzer JXA8900-RL.

Synthetically produced geothermal energy has been used as corrosion medium fluid (North German basin [19], Table 2).

Table 2. Chemical composition of the Northern German Basin (pH=6)

	NaCl	KCl	$CaCl_2$ x $2H_2O$	$MgCl_2$ x $6H_2O$	NH_4Cl	$ZnCl_2$	$SrCl_2$ x $6H_2O$	PbC	Na_2SO_4
g/L	98,22	5,93	207,24	4,18	0,59	0,33	4,72	0,30	0,07

Round corrosion fatigue samples (according to DIN EN ISO 11782-1 and the recommendation of the FKM Research Issue 217) of X2CrNiMoN22-5-3 with shoulder heads were manufactured without additional surface finish as well as polished in the critical region of the specimen. Using a corrosion chamber directly fixed on the test specimen [6-8,14-18]. The corrosion fatigue tests were performed on a Schenck-Erlinger Puls PPV test machine at 33 Hz, 369 K at a flow rate of the medium V*= 2,5× 10^{-6} m³/s ensuring a theoretical flow rate of $\omega_0 = 1{,}7 \times 10^{-3}$ m/s at the critical section of the specimen. To characterize surface corrosion SEM/EDX was performed using: Leo Gemini 1530 VP, acceleration voltage 15 kV. Non-descaled parts of the samples were embedded in a cold resin (Epoxicure, Buehler), cut and polished first with SiC-Paper from 180 µm to 1200 µm under water and then finished with diamond paste 6 µm, 3 µm and 1 µm. Etching was done using different solutions, such as Beraha II. The measurement of the layer thicknesses and phase thickness as well as microstructure analysis were performed via light and electron microscopy techniques using the semi-automatic analyzing program Analysis Docu ax-4 by Aquinto. Additional analysis of the alloy's chemical composition was gathered via Electron Probe Microanalyzer JXA8900-RL separately for each phase.

3. Results and Discussion

Fractographical analysis of the corrosion fatigue cracking was investigated via etched microsections. Figure 2 shows a typical cross section through a highly developed macro crack with significant residual surface opening. This is referred to as an entire major crack region. Generally, the crack path is normal to the direction of applied cyclic push-pull load and therefore predominantly perpendicular to the lamellar duplex phases [17]. Likewise, the crack propagates through austenitic and ferritic phases without any preferred or weakest propagation region.

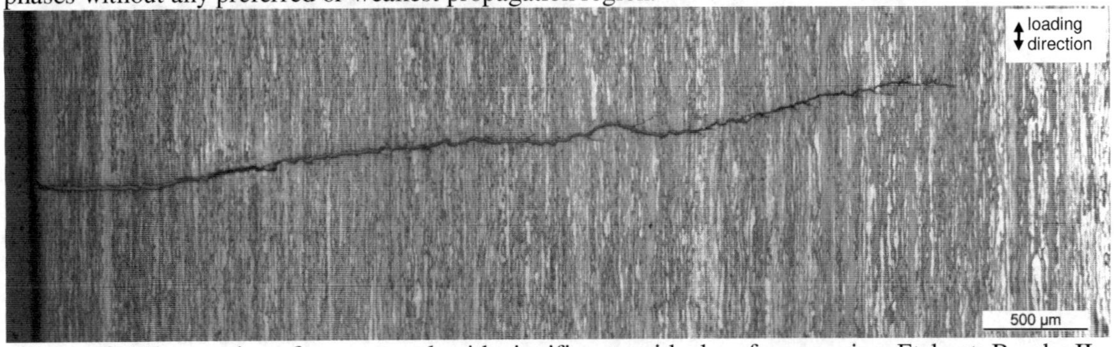

Figure 2. Microsection of macro crack with significant residual surface opening. Etchant: Beraha II.

After initiation cracks propagate strictly linearly with a residual crack width of approximately 30 µm measured within the dedicated crack initiation region attributed to the oldest cavity area. Figure 2,

3 and 4 proof that the austenitic phase (light grey) was susceptible to degradation, while the ferritic phase remained uncorroded leaving straight crack lines. This selective deterioration indicates local electric potential dominance of the ferritic phase [17]. Element distribution measurements revealed significant differences regarding the content of those chemical elements most important for the corrosion resistance, such as molybdenum, chromium and nitrogen. The resulting lower PREN value (Pitting Resistance Equivalent Number) as standard measurement to determine corrosion resistance of stainless steels is 37.39 for the ferritic and 32.37 for the austenitic phase [17]. The lower the PREN the less corrosion resistant is the steel quality. Therefore, the austenitic phase clearly reveals lower corrosion resistance. Another possible explanation is the difference in lattice structure. The face cubic centred structure fcc of the austenitic phase contains large octahedral interstitial lattice voids with a mean diameter of: ca. 0.41 R (R=average atomic radius). On the other hand, the body cubic centred ferrite bcc is characterized by rather small octahedral interstitial lattice voids of: ca. 0.20 µm. Diffusion processes are significantly higher in structures with larger voids leaving austenite high susceptible towards the intrusion of corrosive elements from the brine, such as chlorides. Pitting is therefore initiated at the austenitic phase or within the phase boundary of austenite and ferrite.

Approximately 3 mm below the sample surface the crack propagation path reveals multiple secondary micro cracks (Figure 2). However, these micro cracks remained small in width and soon discontinued after initiation, leaving only a single major propagation path.

Besides corroded crack flanks corrosion pits precipitated that are associated with minor crack, such as micro cracks without significant residual opening (Figure 3). Horizontal corrosion within grains as well as along grain boundaries proceeds beneath the subsurface zone forming new localized corrosion cavities as soon as this distinct area of preferred deterioration has contact to the electrolyte, e.g. the geothermal brine. In addition to corrosion phenomena along the crack flanks (Figure 3) the crack's cavity mainly shows degradation of the austenitic phase which may be of importance considering the initiation of the pits. Due to dissolution of the corrosion pit's region, however, it is not possible to verify this assumption. However, fracture surfaces mainly comprise of the ferritic phase due to early local disappearance of the austenitic phase (Figure 3, right).

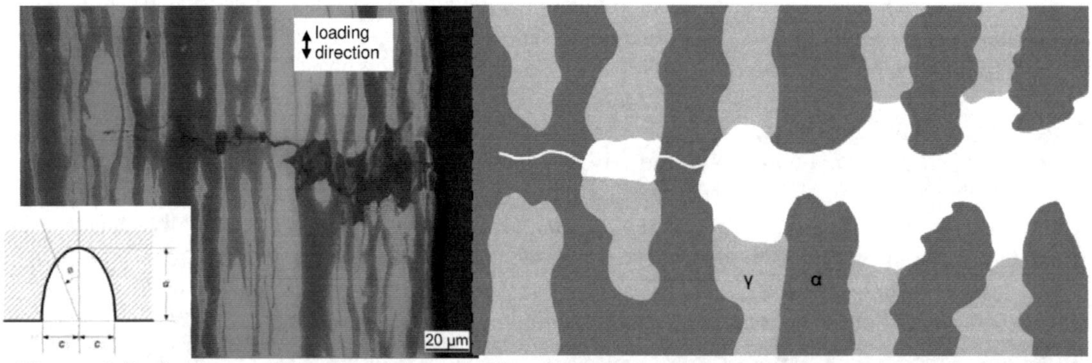

Figure 3. Left: Cross section through corrosion pit. Etchant: Beraha II. Right: schematic drawing showing the preferred deterioration of the austenic phase (γ).

Two possible corrosion fatigue mechanisms are possible, both derived from corrosion fatigue in the passive stat where initial corrosion phenomena act as catalysator of fatigue crack initiation and formatio. The mechanical stress enhances fatigue and crack propagation with corrosive deterioration of the crack flanks as a secondary phenomenon. The first mechanism is derived from the local corrosion effects and the second from degradation of the passivation layer (Figure 4):

1. Pit formation: The CF mechanism may be explained by the initial formation of the corrosion pit, followed by the crack initiation and depassivation of the micro crack area. As a consequence, cracks propagate in the case of sufficient stress being present in the corresponding region leading to early failure of the material.

2. Degradation of passive layer: The CF mechanism may be explained by local destruction of the passivation layer leaving newly formed surface exposed to the surrounding corrosive media. Persistent slip bands formed as result of mechanical axial push-pull load at the surface enhance the surface area where pits are easily formed. Both, exposed surface area and pits increase the degradation of the non-passivated area which then consequently leads to micro crack formation.

Both initiation theories will inevitably lead to crack formation and degradation of the microstructure within the crack flank region (Figure 4 right).

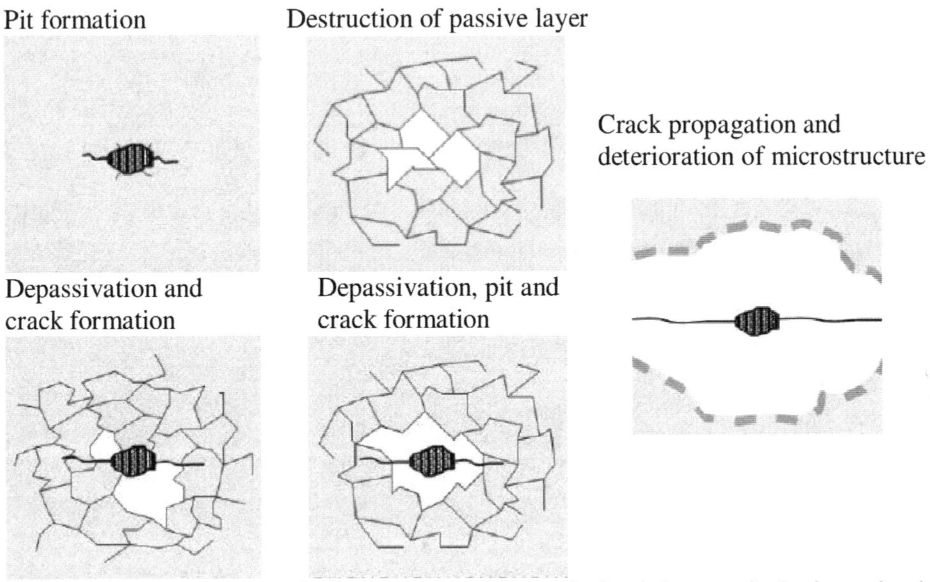

Figure 4. Schematic failure routes during corrosion fatigue. Left: pitting results in depassivation and crack formation. Middle: Depassivation leads to pitting and crack formation. Right: Crack formation and degradation of the microstructure within the crack flank region.

4. Conclusion

To evaluate the failure mechanism, degradation of the steel and crack initiation and propagation the microstructure of different corrosion fatigue samples were investigated via etched micro cross sections. The initial crack region is characterized by a linear path perpendicular to the load applied. Then the crack feathers apart until the crack termination region may be described as a delta-like micro crack structure with an overall curved path. The cavity of immediate subsurface zone is most likely related to the earliest crack cavity exhibiting considerable degradation of the austenitic phase. At the same time the ferritic phase remains undamaged. Crack cavities allocated to later crack formation is generally located deeper underneath the subsurface. Crack initiation may be due to early pit formation resulting in depassivation but also due to local depassivation then resulting in pit formation – both initiation mechanisms lead to crack propagation and early failure. No deterioration of the material is noticed close to the specimen`s axis. The appearance of corrosion pits results in horizontal grain attack of the austenitic phase, assuming the lower corrosion resistance of the γ -phase according to PREN and distinct structural related diffusion processes.

5. References

[1] Thomas J P, Wei R P, *Mater. Sci. Eng.* A 1992 **159** 205–221
[2] Mu L J, Zhao W Z 2010 *Corros. Sci.* **52** 82–89
[3] Unigovski Y B, Lothongkum G, Gutman E M, Alush D, Cohen R, 2009 *Corros. Sci.* **51** 3014–3020
[4] Holtam C M, Baxter D P, Ashcroft I A, Thomson R C 2010 *Int. J. Fatigue* **32** 288–296
[5] Thorbjörnsson I 1995 *Mater. Des.* **16** 97–102

[6] Pfennig A, Wolf M, Wiegand R, Kranzmann A, Bork C-P 2013 *Energy Procedia* **37** 5764–5772
[7] Pfennig A, Wiegand R, Wolf M, Kranzmann A, Bork C-P 2013 *Corros. Sci.* **68** 134–143
[8] Pfennig A, Wolf M, Heynert K., Böllinghaus T 2014 *Energy Procedia* **63** 5773–5786
[9] Ebara R, 2006 Eng. Fail. Anal. **13** 516–525
[10] Alvarez-Armas I 2008 Mech. Eng. **1** 51–57
[11] Prosek T et al 2014 *Corrosion* **70** 1052–1063
[12] Schultze S, Göllner J., Eick K, Veit P, Heyse H 2001 *Mater. Corros.* **52** 26–36
[13] Arnold N, Gümpel P, Heitz T W 1999 *Mater. Corros.* **49** 140–145
[14] Pfennig A, Trenner S, Wolf M, Bork C-P 2013 in European Corrosion Congress EuroCorr, September 1 - 5, Estoril Congress Center , Estoril, Portugal
[15] Wolf M et al 2017 *Energy Procedia 114* 5337 – 5345
[16] Pfennig A, Gröber A, Simkin R, Kranzmann A 2019 *Matter: Int. J. Sci. and Tech.* **5** (1) 609-631
[17] Pfennig A, Simkin R, Gröber A, Kranzmann A 2019 Green House Gas Emission Red. Techn. (GHGT14) Melbourne, Australia, 21st-26th October 2018
[18] Pfennig A, Kranzmann A 2019 2nd International Conference on Materials Engineering and Applications (ICMEA 2019), January 16-18, 2019 Da Nang, Vietnam
[19] Förster A et al 2010 *Mar. Pet. Geol.* **27** 2156–2172

IOP Conf. Series: Materials Science and Engineering **894** (2020) 012016 doi:10.1088/1757-899X/894/1/012016

Effects of Texture on the High Temperature Scratch Wear Behavior in Zinc

Aditya Gokhale[1], E-Wen Huang[2], Soo Yeol Lee[3], Rajesh Prasad[1], Jayant Jain[1*]

[1]Department of Materials Science and Engineering, Indian Institute of Technology Delhi, New Delhi-110016, India.
[2]Department of Materials Science and Engineering, National Chiao Tung University,1001 University Road, Hsinchu, Taiwan 30010, Taiwan (R.O.C.).
[3]Department of Materials Science and Engineering, Chungnam National University, Daejeon 305-764, Republic of Korea

jainjayan@gmail.com

Abstract: In this study the effect of texture on the wear behavior was investigated in extruded rod of pure zinc. Different loads and temperatures were used to investigate the effect of texture on wear behavior. Formation of chips at higher temperature marked the transition of wear mechanism from micro-ploughing to micro-cutting. Suppression of twinning at high temperature was also observed suggesting a complete slip dominated deformation at high temperature. The effect of texture was evident at high temperature and high loads.

1. Introduction

Wear is an omnipresent phenomenon in engineering service life of a product. From machining to micro-fabrication the component undergoes wear in some way or the other. However, in some cases, like precision machining, the wear and presence of surface damage becomes critical in deciding the performance of the material/component. Besides being used as a coating, zinc is also used in small electrical housing and medical implants [1, 2]. Most of the wear studies are centered around the wear endurance by conducting the pin on disc tests bringing out the bulk wear response. However, microstructural and texture aspects are often overlooked. Scratch studies offer better control to study the microstructural variables and scale dependent response. Parameters like grain size, microalloying and grain orientations have found to influence the wear resistance. Mezlini et al. [3] investigated the effect of grain size on the scratch wear behavior in Al-alloy and found out that smaller grain sizes can improve the wear resistance. Shishir et al. [4] investigated the effect of Zn addition in Mg on the wear behavior of Mg-Zn binary alloy and found that addition of Zn increased the wear resistance. Pranjal et al. [5] have shown that aging of the AZ80 alloy improves the scratch wear resistance. Shugurov et al. [6] investigated ploughing as a function of grain orientation in Ti. Swadener et al. [7] studied the effect of grain orientation and scratch wear in grains of Ti-Nb alloy and observed that work hardening due to scratch had a strong crystallographic correlation. Csanádi et al. [8] investigated the nano-scratch response in ZrB_2 crystals and found that grains having basal orientation was the hardest grain showing highest scratch resistance. Flom and Komanduri [9] performed sliding experiments on different materials including single crystal zinc, however, they did not track the wear mechanism and associated deformation mechanisms. Somekawa et al. [10] investigated small volume wear in single crystal pure Mg. At higher loads the scratch on basal plane resulted in more wear when compared to prismatic plane.

The wear rate was also found to be governed by the prevalent deformation mechanism. Motivated by the fact that harder will resist better, the present work demonstrates the effect of texture and temperature on the scratch wear behavior in pure zinc.

2.Experimental procedure

Extruded rod of pure Zn (99.99%) was obtained from Alfa Aesar (Thermo Fischer Scientific India Pvt. Ltd). Two set of samples were obtained such that the surface to be polished and scratched was perpendicular to the extrusion direction (ED) and in the other the surface was parallel to the ED. The specimen was ground using SiC, followed by cloth polishing in a colloidal silica medium until mirror finish was obtained. The sample was etched using 2% Nital for 8 s to reveal the grains. ASMEC's Universal Nanomechanical Tester (Bautzner Landstraße 45, Germany) was used to perform scratch. A diamond cono-spherical tip, with tip radius ~10 µm was used with a scratch velocity of 5 µm/s. Room temperature scratch test was carried out at 22°C whereas high temperature testing was carried out at 100°C and 150°C on the two different sets of samples which have different texture. For every temperature, a new sample was used so that the effect from previous heating is eliminated. The sample was placed on the heating plate to ensure direct contact. The scratch test progressed after an initial time of 180 s was provided in order to stabilize the system and ensure the tip is heated. Hitachi's back-scattered electrons-based tabletop scanning electron microscope (SEM) TM3000 (Tokyo, Japan) was used to observe the nearby areas of the scratch. Macwin Zeta 20 optical profilometer was used to obtain the scratch groove cross section area. For each condition 3 scratches were made, and 4 cross sections were taken making it 12 cross sections.

3.Results and Discussion

The extruded rod of zinc exhibits a basal texture with the majority of the basal planes parallel to the ED [11]. The scratches were performed by scratching in the direction of ED and perpendicular to ED.

3.1 Wear mechanisms

Vlassak and co-worker [12] investigated the change in hardness as a function of orientation in single crystal zinc. It was found that for near basal orientation the hardness was ~0.64 GPa, whereas for orientation perpendicular to $< c >$ axis gives a value of ~0.56 GPa. For a constant load, the wear rate should become a function of temperature wherein the wear rate will increase as the temperature increases. The increase in hardness is 14%, but a marked difference in the wear rate can be seen for two case, shown in Figure 1. The wear rate measured for scratch parallel to ED was lower as compared to scratch perpendicular to ED. At room temperature the wear rate is nearly same for 50 mN. However, as the temperature is increased, one can observe the effect of texture. Considering the Archard equation (eq. 1) for abrasive wear one can relate hardness and wear rate by [13]:

$$\frac{Q}{L} = K\frac{P}{H} \tag{1}$$

where Q is the wear volume, L is the length of scratch, P is the load, H is the hardness and K is a constant. From eqn. (1), as the slope increases the wear rate increases as a function of temperature, characterized by the change in slope. At higher loads and temperatures, however, the effect of texture is pronounced. At 200°C the effect of texture become evident. For 100 mN and 150 mN the slope does not get affected much. There is a glaring effect of texture and the surface perpendicular to ED shows more wear rate, even at lower loads. Figure 2 shows the wear tracks for a surface parallel to the ED and perpendicular to ED, scratched at different loads and temperature. At the outset, for scratches made parallel to ED, one can observe the chip formation at high temperature. At room temperature micro-ploughing dominated at all loads with lack of chip formation in both the cases. Under large negative rake angle, a brittle material can also develop a continuous chip while machining [14]. Zinc has low ductility and continuous chip formation is absent while machining a low ductile material as continuous chip formation is a signature for ductile material. The high negative rake angle is capable

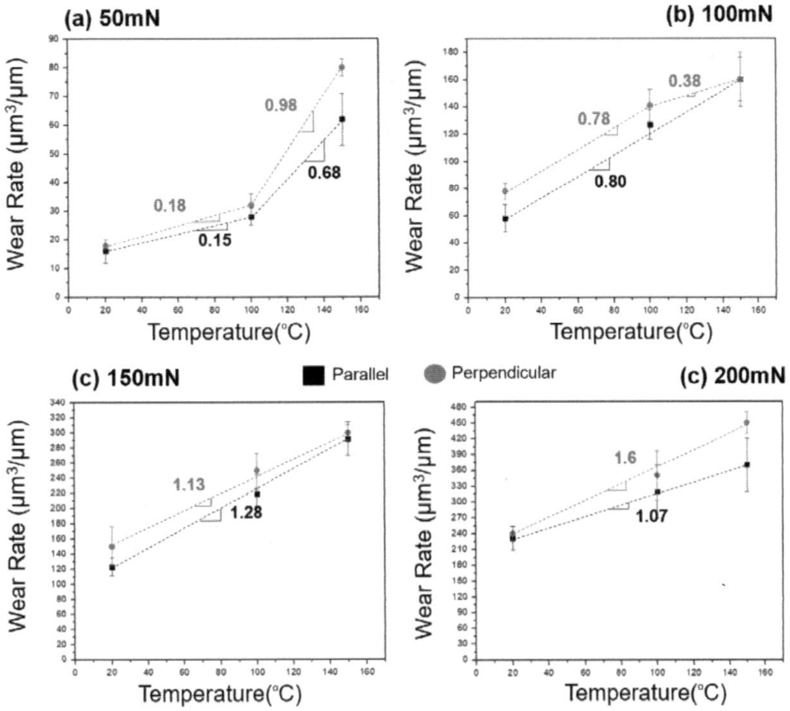

Figure 1. Wear rate at different temperatures for different orientations.

of developing a high hydrostatic pressure under and ahead of the tool thereby forming a plastic zone. Chip debris were observed on the sides of the scratch groove for both the cases at room temperature. The change in wear mechanism was observed as the temperature changed. For scratches along ED, as the temperature increased the tendency for chip formation also increased, however for scratches perpendicular to ED the chip formation could not be sustained. For scratches perpendicular to ED chip breakage was also more prominent in this orientation.

3.2 Deformation mechanisms

Figure 3 (a-f) and (g-l) shows the deformation mechanisms near the scratched groove for surface parallel to the ED and perpendicular to ED, respectively. Owing to the texture, scratches made perpendicular to ED effectively compresses the crystals in $< c >$ axis. Similarly, the scratch made parallel to ED puts the $< c >$ axis in extension. However, the motion of indenter thus induces stress in different directions activating other deformation modes. Slip was observed to be the major deformation modes associated with the scratch. Besides slip, zinc is known to undergo twinning under compression [15, 16]. Twinning phenomenon progress in steps of nucleation and growth. The twins surrounding the scratch grooves look like the vertebrae of a fish, shown in Figure 2 (j). The twin mode observed is of the type $\{10\bar{1}2\} < 10\bar{1}1 >$ compressive twin. As the scratch progresses the twin is incubated, and further movement of indenter induces more compressive strain in the grain thereby a need for the twin to grow. As the indenter progresses ahead the stress level increases but as soon the indenter progresses, there is not enough stress to cause further growth of the twin. The adjacent region undergoes the same cycle and thereby we see multiple twins of the same type developing in the same grains. Plastic deformation of the material is dependent on the number active slip systems and twin systems. The triaxial nature of stress under the indenter is capable of activating basal slip which is the easiest slip system in zinc. Thus, slip lines are distinctly observed on the surface. Another notable

IOP Conf. Series: Materials Science and Engineering **894** (2020) 012016 doi:10.1088/1757-899X/894/1/012016

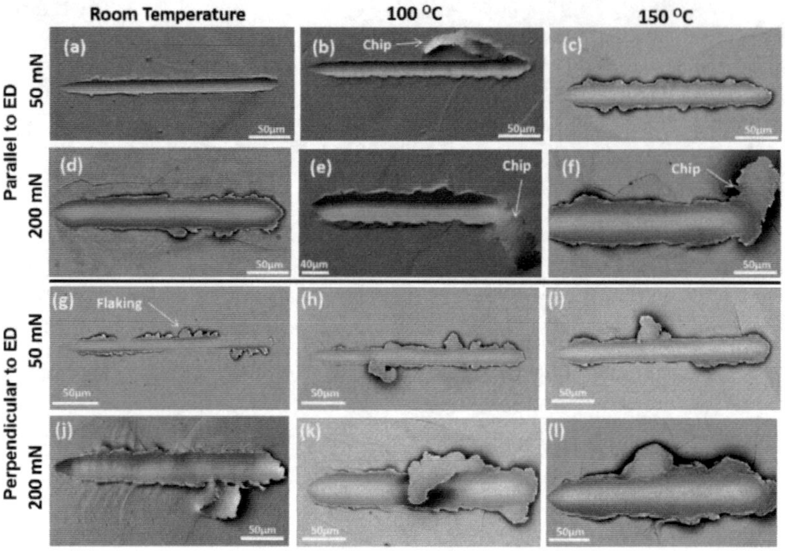

Figure 2. Wear track of scratches made, at different temperatures, (a-f) parallel to ED and (g-l) perpendicular to ED.

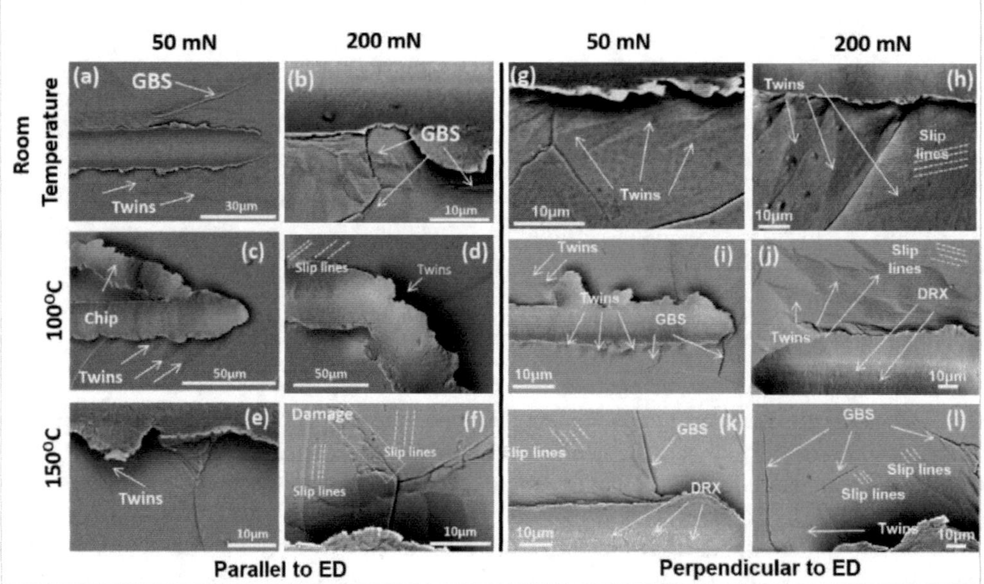

Figure 3. Wear track showing deformation mechanisms for scratches, made at different temperatures, (a-f) parallel to ED and (g-l) perpendicular to ED.

feature is the grain boundary sliding (GBS). The grains have slid past one another giving rise to a ledge. GBS can act as sites for crack generation. At higher temperatures the grain boundaries (GB) are undergoing extensive damage. This points out to the fact that not only a stress is required for incubation but in order for the twin to grow there must be a continued impingement of stress or stress field in order to support the twin growth. The twinning activity has decreased as the temperature has increased. For scratches made at high temperature, the dynamic recrystallization was observed inside the scratch groove, Figure 3 (k). The recrystallized grains are mostly pyramidal, and the twinned region is an active site for recrystallization [11].

IOP Conf. Series: Materials Science and Engineering **894** (2020) 012016 doi:10.1088/1757-899X/894/1/012016

4.Conclusion

The present was able to demonstrate and bring out the influence of texture on the high temperature wear response in zinc. Scratch wear tests was performed on surfaces parallel and perpendicular to ED. For scratches made parallel to ED the chip formation was pronounced than the other case. Wear resistance for scratches made parallel to ED was more. At high temperature and load, the effect becomes more pronounced and surface parallel to ED gives a better wear resistance. Slip was found to be the major deformation mechanism however suppression of twin was observed at high temperature besides GBS.

5. References

[1] Interzinc, Designing Zinc Castings for Electronic Applications, 1994. https://www.eazall.com/PublicDoc/Designing_for_Electronic_Applications.pdf.
[2] P.K. Bowen, E.R. Shearier, S. Zhao, R.J. Guillory, 2nd, F. Zhao, J. Goldman, J.W. Drelich, Advanced healthcare materials 5(10) (2016) 1121-40.
[3] S. Mezlini, M. Zidi, P. Kapsa, Industrial Lubrication and Tribology 60(5) (2008) 255-258.
[4] S. Keerti, A. Gokhale, J. Jain, E.W. Huang, Tribology Letters 65(4) (2017).
[5] P. Nautiyal, J. Jain, A. Agarwal, Tribology Letters 61(3) (2016).
[6] A. Shugurov, A. Panin, A. Dmitriev, A. Nikonov, Wear 408-409 (2018) 214-221.
[7] J.G. Swadener, H. Bögershausen, B. Sander, D. Raabe, Journal of Materials Research 25(05) (2011) 921-926.
[8] T. Csanádi, A. Kovalčíková, J. Dusza, International Journal of Refractory Metals and Hard Materials 65 (2017) 45-51.
[9] D.G. Flom, R. Komanduri, Wear 252 (2002) 30.
[10] H. Somekawa, K. Tomita, M. Toda, T. Hirayama, T. Matsuoka, Metallurgical and Materials Transactions A 46(9) (2015) 4351-4359.
[11] A. Gokhale, R. Sarvesha, R. Prasad, S.S. Singh, J. Jain, Materials Letters 247 (2019) 151-154.
[12] J.J. Vlassak, W.D. Nix, Journal of the Mechanics and Physics of Solids 42(8) (1994) 23.
[13] J.F. Archard, Journal of Applied Physics 24(8) (1953) 981-988.
[14] R. Komanduri, N. Chandrasekaran, L.M. Raff, Wear 219(1) 8.
[15] S. Yoshida, N. Nagata, Transactions of the Japan Institute of Metals 9(2) (1968) 110-119.
[16] W.F. Hosford, Mechanical Behavior of Materials, Second ed., Cambridge University Press, 2010.

IOP Conf. Series: Materials Science and Engineering **894** (2020) 012017 doi:10.1088/1757-899X/894/1/012017

Experiment and Numerical Modeling of Boron Effect on Weld HAZ Hardenability of Steels

K.Yoshimoto[1, a],T.Pasang[2, b],S.Aihara[3, c] and T.Kawabata[4, d]

[1]Department of Systems Innovation, Undergraduate School of Engineering, The University of Tokyo
[2]Head of Department Mechanical Engineering Director-Engineering Research Institute
[3]Department of Systems Innovation, Graduate School of Engineering, The University of Tokyo
[4]Department of Systems Innovation, Graduate School of Engineering, The University of Tokyo

[a]yoshimoto@fract.t.u-tokyo.ac.jp , [b]pasang@aut.ac.nz, [c]aihara@fract.t.u-tokyo.ac.jp, [d]kawabata@fract.t.u-tokyo.ac.jp

Abstract. Boron is an alloying element used for increasing the strength of steels, but on the other hand, it sometimes deteriorates heat affected zone (HAZ) weldability and toughness. Although it is widely known that the hardenability of steel is improved due to grain boundary segregation of boron, its details and quantification have not yet been clarified. In this study, boron-free HT590 and boron-added HT780 steels were welded and microstructure and hardness distribution in HAZ were measured, in detail. Furthermore, we calculated and investigated the grain boundary segregation and diffusion of boron during weld thermal cycle by numerical analysis while comparing it with the experiment result. As a result, the present model made it clear that the effect of boron on HAZ hardenability can be expressed by non-equilibrium segregated boron at grain boundary, quantitatively.

1. Introduction

Boron (B) is a micro alloy element which increases hardenability of steels and it has been widely utilized in steel industry since many years ago [1]. Only a small addition of B, as low as 10ppm, has a tremendous effect on the hardenability of steels [2, 3]. It is widely accepted that B atoms are segregated at austenite grain boundary (GB) and reduce GB energy, thereby retarding austenite to ferrite, bainite and martensite transformation [4-11]. Relationship between concentration of B at G.B. and hardenability has been examined in a wide variety of steel chemical compositions and thermal histories, including isothermal and continuous cooling, intensively. During cooling process Fe-(C,B) precipitation may take place, in which case hardenability is reduced. It is also known that the addition of Molybdenum retards the precipitation, thereby maintaining the B hardenability effect [5, 10, 11]. The boron effect on steels' hardenability is not limited to heat treatment of steels but also has an influence on phase transformation of weld heat-affected zone (HAZ) [12, 13]. In the latter case, HAZ hardness is elevated in B-added steels and sometimes promotes HAZ cold cracking and reduction in HAZ toughness. Generally, B has a deleterious effect on HAZ weldability.

As mentioned above, B increases HAZ hardenability but in somewhat unusual manner. In B-free steels, maximum hardness is realized at coarse-grained (CG) HAZ near fusion line (FL) and the hardness decreases with increasing distance from the FL, along with lowered peak temperature of weld thermal cycle and reduced austenite grain size. However in B-added steels, maximum hardness often appears in the HAZ well apart from the FL, where the peak temperature is lower than that at CG HAZ [12, 13]. It is argued that the B segregation at GB is enhanced at lower peak temperature due to non-equilibrium segregation [13]. But concrete mechanism has not been established, yet.

In the present study, weld HAZ hardenability was investigated in B-free and B-added steel. Furthermore, numerical simulation on B diffusion and segregation was conducted and the influence of peak temperature and cooling rate on B hardenability are discussed.

2. Weld experiment

Welding was conducted to examine the B hardenability effect in HAZ.

2.1. *Steels tested*

Table 1 shows chemical compositions of the steel plates tested in the welding. They are as-rolled HT590 and quenched-tempered HT780 commercial steels. The HT590 steel contains practically no B and the HT780 steel contains 12ppm B. Plate thickness was 19mm in the both steels.

Table 1. Chemical compositions of steels for welding.

Steel 1	**Chemical compositions(mass%)**										**Plate thickness(mm)**
	C	Si	Mn	P	S	Cr	Ti	Al(sol)	B[ppm]	N[ppm]	
H590	0.10	0.24	1.70	0.017	0.005	0.02	0.091	0.033	1	30	19
H780	0.15	0.23	1.28	0.012	0.006	0.25	0.012	0.025	12	37	19

2.2. *Welding conditions*

Submerged-arc welding was conducted. Table 2 shows the welding conditions, Heat input was 40kJ/cm. Weld bead was placed on the steel plate, i.e., bead-on-plate welding. A section sample was cut from the weld, polished and etched by 2% nital solution to reveal microstructure. Vickers hardness was measured traversing weld metal (WM), FL, CG HAZ, fine-grained (FG) HAZ, inter critical (IC) HAZ and base metal (BM) with indentation force of 0.5N.

Table 2. Welding conditions.

Wire	**Flux**	**Wire diameter (mm)**	**Current(A)**	**Voltage(V)**	**Speed(cm/min)**	**Heat input(KJ/cm)**
Linconweld L-61	Lincolnweld 860	2.4	500	50	37	40

2.3. *Result*

Figure 1 shows microstructure distributions with Vickers hardness indentations for the HT590 and HT780 welds. Figure 2 shows hardness distributions of the both welds. The HT590 weld shows typical hardness profile; maximum hardness appeared at CG HAZ near the FL and decreased with increasing distance from the FL. Contrary to the HT590 weld, the HT780 weld showed nearly flat hardness profile from the FL to FG HAZ but with considerable hardness drop at inter critical HAZ. It should be noted that the HT590 steel contains 0.091% Ti, which might have increased overall HAZ hardness by Ti carbide precipitation after phase transformation. Figure 3 shows microstructures at WM, CG HAZ, FG HAZ, IC HAZ and BM. The both welds showed columnar structure at WM, and predominantly bainitic microstructure in HAZ and ferrite-pearlite (HT590) and martensitic microstructures (HT780) at BM.

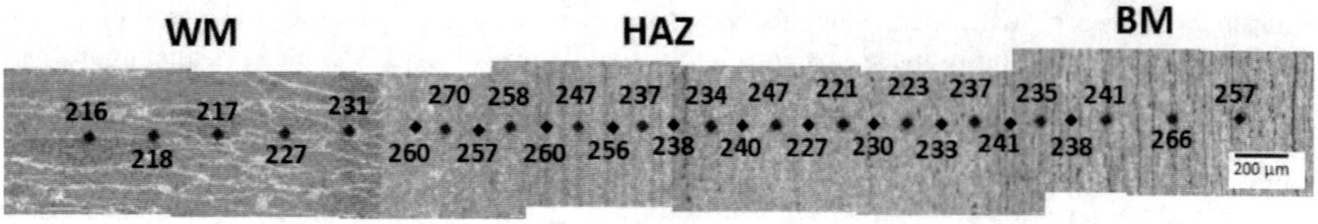

(a) HT590 weld

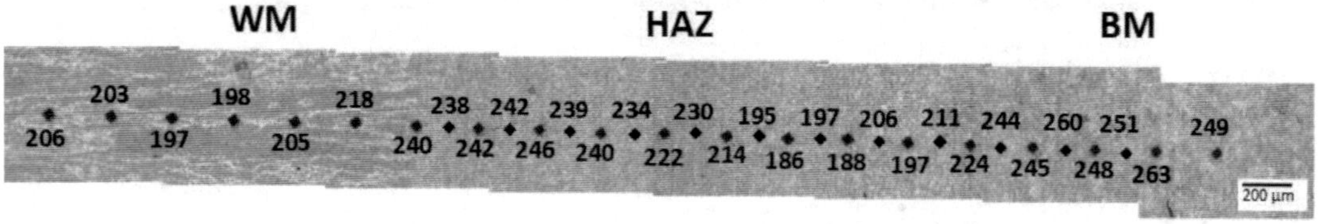

(b) HT780 weld

Figure 1. Microstructure distributions in weld.

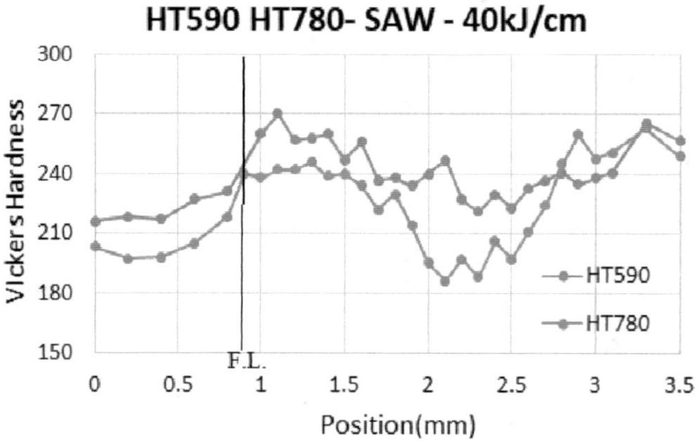

Figure 2. Hardness distributions of welds.

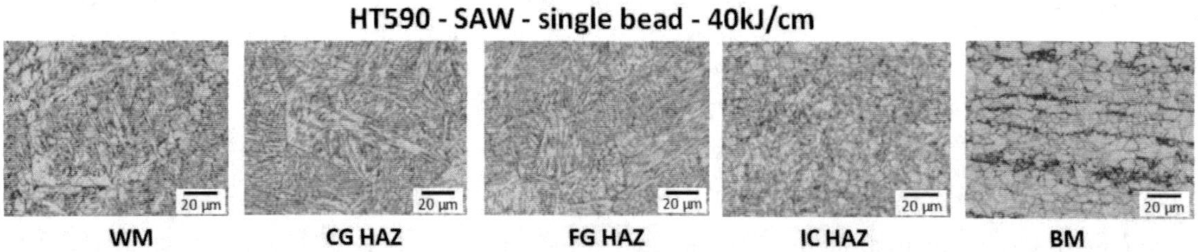

(a) HT590 weld

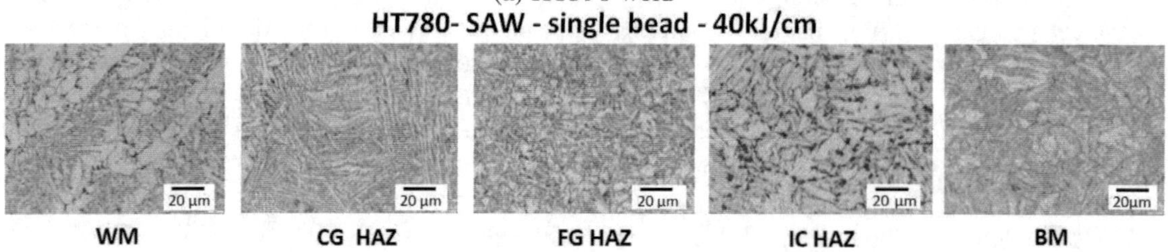

(b) HT780 weld

Figure 3. Microstructures in weld.

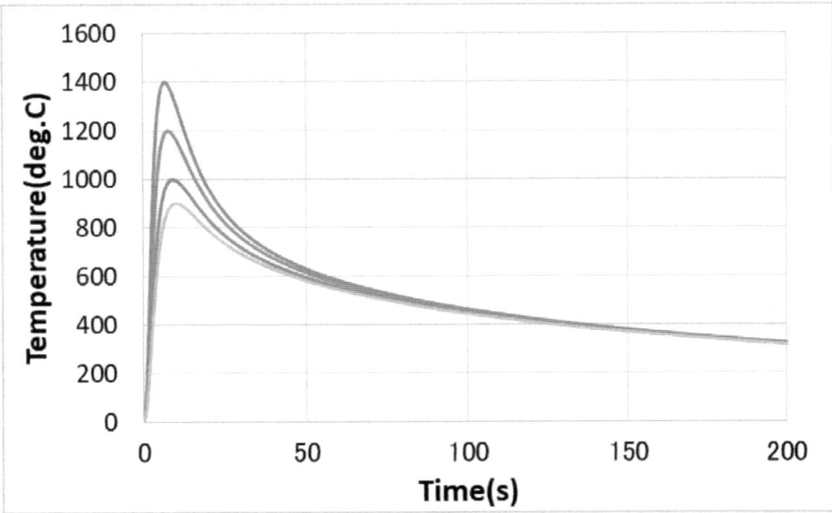

Figure 4. Calculated thermal history.

It should be noted that average cooling rate between 800℃ and 500℃ was calculated as about 5 ℃/s using in-house software of weld heat conduction analysis, see figure 4. The hardness profile shown in figure 2 indicates that HAZ hardenability is maintained in the HT780 weld even if the peak temperature of weld HAZ thermal cycle is low, as compared with the hardenability drop from CG to FG HAZ in the HT590 steel.

3. Numerical analysis of boron diffusion and GB segregation

For the purpose of understanding the possible change of the boron hardenability effect with thermal cycle conditions, segregation of B at GB was calculated in conjunction with B diffusion in austenite, numerically

3.1. Procedure of numerical analysis

The theory of equilibrium segregation of impurity elements was established by Mclean [14]. The present analysis follows his theory. Eq.(1) gives concentration of B at GB, $C_{B(GB)}$.

$$C_{B(GB)} = \frac{C_B exp(-Q_B/RT)}{1 - C_B + C_B exp(-Q_B/RT)} \tag{1}$$

where C_B is concentration of B in austenite, Q_B is activation energy, R is universal gas constant and T is temperature. This equation may be applied if temperature is kept for sufficiently long period. However, the sample is continuously cooled from peak temperature down to phase transformation temperature, in which case non-equilibrium segregation must be considered in conjunction with B diffusion in austenite grain, see figure 5. Coupled analysis of the non-equilibrium segregation and diffusion of B in austenite was conducted numerically in the present study. It is assumed that equilibrium distribution of B is achieved at peak temperature because peak temperature is relatively high and diffusion is fast. Upon cooling, local equilibrium between GB and austenite just adjacent to the GB can be assumed, in which case C_B in Eq.(1) is regarded as B concentration at $x = 0$ (austenite just adjacent to the GB). Assuming one-dimensional diffusion, B diffusion in austenite grain is expressed as,

$$\frac{\partial C_B}{\partial t} = D_B \frac{\partial^2 C_B}{\partial x^2} \tag{2}$$

where D_B is diffusion constant of B in austenite grain and a function of temperature. Flux of B atoms from austenite to GB is calculated as,

$$J_{B(GB)} = -D_B \frac{\partial C_B}{\partial x}|_{x=0} \tag{3}$$

Increase in $C_{B(GB)}$ during time increment can be calculated as,

$$dC_{B(GB)} = \frac{J_{B(GB)}}{w_{GB}/2} dt \tag{4}$$

where w_{GB} is GB width. C_B and $C_{B(GB)}$ can be calculated numerically along the thermal cycle.
1: Set initial value of $C_{B(x)}$ (uniform in austenite grain) and $C_{B(GB)}$ at initial (peak) temperature.
2: Determine temperature at next time step, following the assumed weld thermal cycle.
3: Determine $C_{B(x=0)}$ in local equilibrium with $C_{B(GB)}$.
4: Solve diffusion equation using finite-difference algorithm and obtain $C_{B(x)}$.
5: Calculate $C_{B(GB)}$ using $dC_{B(GB)}$.
6: Repeat 2 to 5.

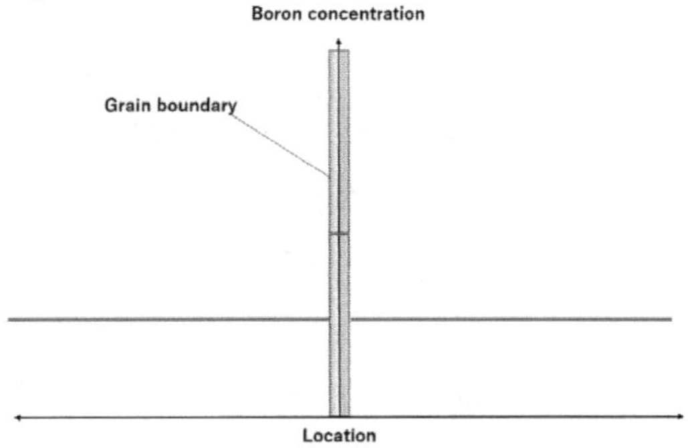

(a) Initial state

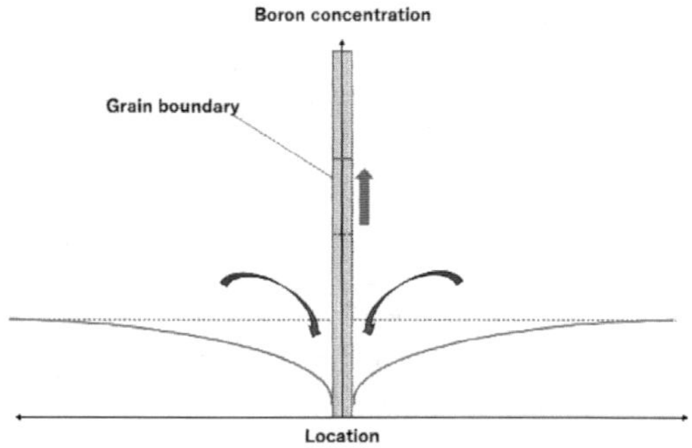

(b) After boron diffusion

Figure 5. Schematic diagram of boron diffusion.

3.2. Results of numerical calculations

Figure 6 shows change of boron concentration with time for peak temperature:1000°C,1400 °C, where the cooling curves of figure 4 were applied. It can be seen that at lower peak temperature B diffuses faster to the GB in spite of the lower diffusion constant. This is due to the local equilibrium bwetween GB B and that in grain adjacent to the GB. Figure 7 shows change of GB B concentration with time for different peak temparatures, but plotted against temparature during cooling. The GB B concentration increases with decreasing temparature during cooling but its value became higher for the lower peak temperature. Figure 8 shows GB B concentration at 650°C as a function of peak temperature. It is clearly seen that GB B concentration was higher with lower peak temperature, especially at 900°C.

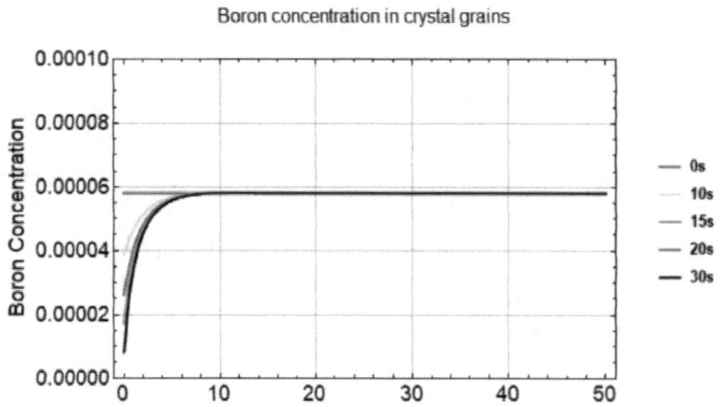

(a) Peak temperature:1000°C

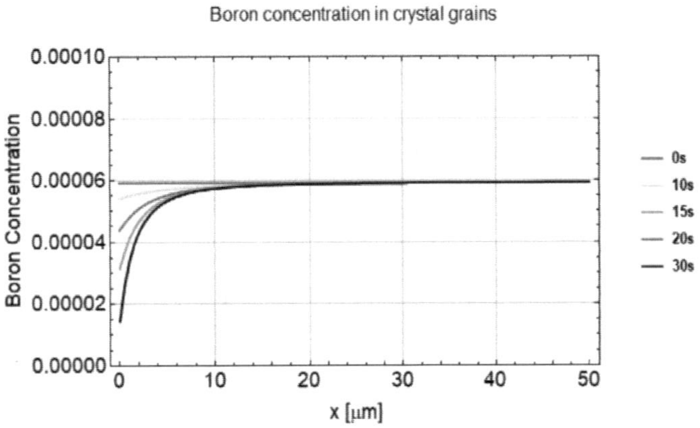

(b) Peak temperature:1400°C

Figure 6. Calculated Boron concentration in crystal grains.

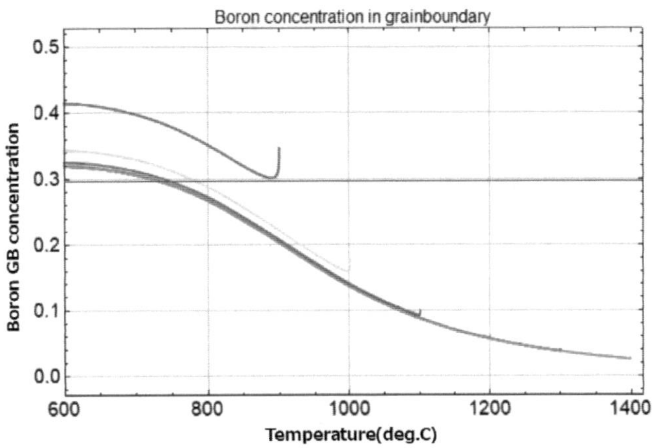

Figure 7. Calculated Boron grain boundary segregation concentration.

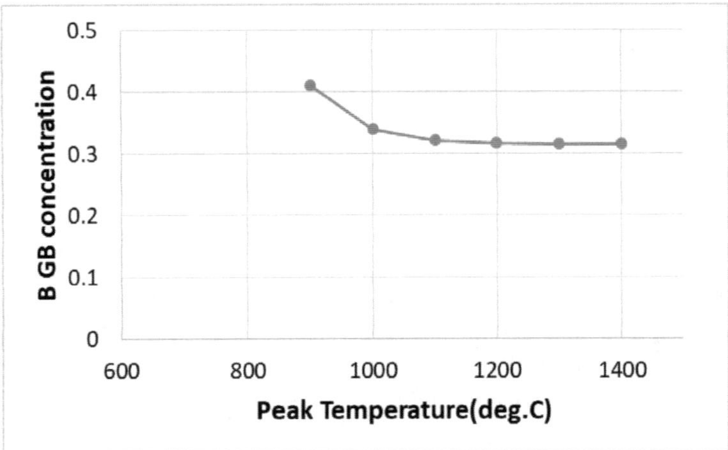

Figure 8. Calculated peak heating temperature dependence of GB segregated boron concentration at 650°C.

3.3. Discussions

Although the hardness distribution of weld HAZ HT780 tended to be flattened than HT 590, the change in the hardness distribution by B was not necessarily clear in the present experiment. It is known that GB B is sometimes precipitated as $Fe_{23}(CB)_6$ in the course of cooling. In this case, the effect of B on hardenability is reduced[2, 7, 11]. On the other hand, Mo is known to suppress the B precipitation [10, 11]. The HT780 steel in the present experiment did not contain Mo. Therefore, results of HAZ hardness measurement of B added steel containing Mo by Kasuya et al [12] are shown in figure 9 and figure 10, together with B-free steel. Figure 9 shows schematic illustration of hardness distribution. While the B-free steel had hardness distribution similar to the present experiment, see figure 10(a), highest hardness is realized at a position away from FL in the Mo-B steel, see figure 10(b). This result corresponds to the dependence of GB B concentration on peak temperature, see figure 8. That is, the GB B concentration is high when the peak temperature is 1000°C or lower. This result suggests that the effect of B on HAZ hardenability can be explained by GB B concentration. However, it is necessary to consider the influence of B precipitation at the same time for Mo free B added steels.

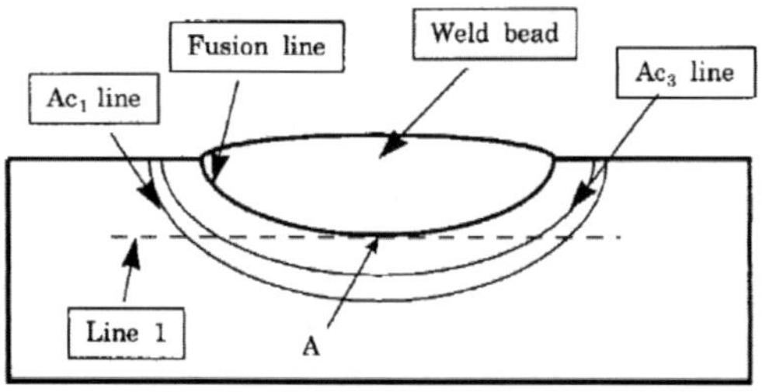

Figure 9. Schematic illustration of hardness distribution[12].

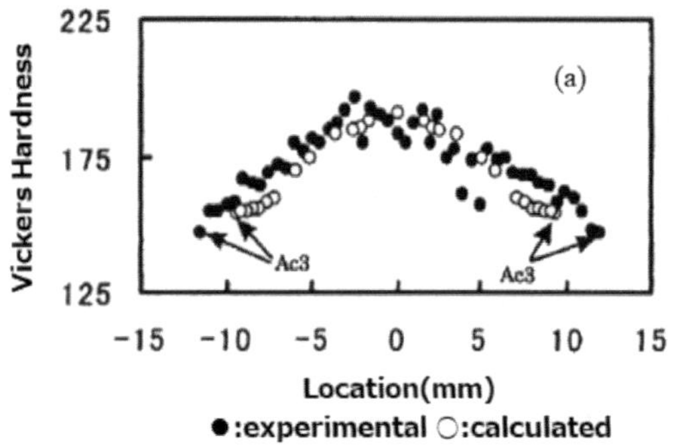

(a) B-free steel, 48kJ/cm

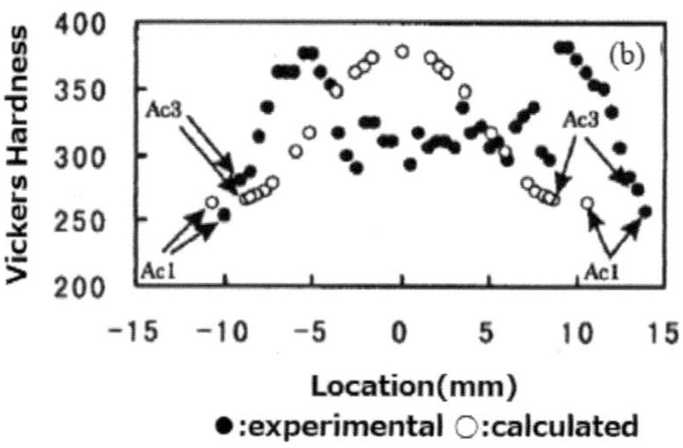

(b) Mo-B steel (21Bppm), 48kJ/cm

Figure 10. Hardness distribution of the steel [12].

4. Conclusions

(1) It has been made clear from the present experiment together with the published data that weld HAZ hardness distribution changes with the addition of B.

(2) The present numerical simulation on the diffusion and grain boundary segregation of B showed a change of GB B concentration on peak temperature of weld HAZ thermal cycle.

(3) It is suggested that HAZ hardenability of B added steels can be explained by grain boundary B concentration.

In the future, it is necessary to analyze the grain boundary segregation of B considering B precipitation.

5. References

[1] C.R. Simcoe, A.R. Elsea and G.K. Manning, Further work on the boron-hardenablity mechanism, J. Metals, Transactions AIME, 984-988, August 1956.

[2] M. Ueno, Distribution of boron at austenite grain boundaries and bainitic transformation in low carbon steels, Trans. ISIJ, vol.13,1973, 211-217.

[3] H.K.D.H. Bhadeshia, Bainite in Steels, 183, The Institure of Materials, 1992.

[4] L. Karlsson, H. Norden and H. Odelius, Non-equilibrium grain boudary segregation of boron in austenitic stailess steel -I, Acta metall. Vol.36, No.1, 1-12, 1988.

[5] L. Karlsson and H. Norden, Non-equilibrium grain boudary segregation of boron in austenitic stailess steel -II, Acta metall. Vol.36, No.1, 13-24, 1988.

[6] L. Karlsson, Non-equilibrium grain boudary segregation of boron in austenitic stailess steel -III, Acta metall. Vol.36, No.1, 25-34, 1988.

[7] L. Karlsson, Non-equilibrium grain boudary segregation of boron in austenitic stailess steel -IV, Acta metall. Vol.36, No.1, 35-48, 1988.

[8] X.L. He, Y.Y. Chu and J.J. Jonas, Grain boudary segregation of boron furing continuos cooling, Acta metall. Vol. 37, No.1, 147-161, 1989.

[9] X.L. He, Y.Y. Chu and J.J. Jonas, The grain boundary segregation of boron furing isothermal holding, Acta metall. Vol. 37, No.11, 2905-2916, 1989.

[10] H. Asahi, Effects of Mo addition and austenitizing temperature on hardenability of low ally V-added steels, ISIJ Int., vol.42 (2002), No.10, 1150-1155.

[11] J. Takahashi, K. Ishikawa, K. Kawakami, M. Fujioka and N. Kubota, Atomic-scale study on segregation behavior at austenite grain boundaries in boron- and molybdenum-added steels, Acta Materiala 133 (2017) 41-54.

[12] T. Kasuya and Y. Hashiba, Prediction of hardness distribution in steel heat-affected zone, Science oand Technology of Welding and Joining, 1999, vol.4, No.5, 265-275.

[13] T. Kamo, M. Hamada and Y. Komizo, Hardenss distribution in weld heat affected zone of boron bearing low carbon steel, Bulletin of Japan Welding Society, vol.20, No.2, 276-281 (2002).

[14] D. McLean, Grain Boundaries in Metals, 116, Oxform at the Clarendon Press, 1957.

IOP Conf. Series: Materials Science and Engineering **894** (2020) 012018 doi:10.1088/1757-899X/894/1/012018

Effect of Segment Thickness on Internal Force of Shield Tunnel under Abnormal Loading Conditions

Bolin ZHANG[1], Jun ZHENG[1*]
[1]Nanjing Metro Group Co., Ltd., Nanjing 210018

E-mail:1120319381@qq.com

Abstract. Segment thickness is one of the important parameters of shield tunnel structure design. Segment thickness determines the section stiffness and mechanical performance of shield tunnel structure. Considering the load type of overload and unload, a three-dimensional fine shield tunnel numerical model was established. The full-scale test was used to verify the correctness of the numerical model. In the end, the effect of segment thickness on the internal force of shield tunnel structure under abnormal loading conditions was analyzed. The results show that the internal force of each section of the tunnel structure increases with the increase of the thickness of the segment, and the difference of the internal force between the different segment thickness increases with the increase of the overload and the degree of unloading. Increasing the thickness of segment will delay the breakpoint of the bending moment and axial force (critical instability state). The excessive increase of the segment thickness can lead to a large increase in the internal force of the tunnel structure, resulting in a decrease in the crack resistance of the tunnel structure.

1. Introduction

The relative stiffness of shield tunnel structure and surrounding soil affects the interaction between them, which changes the internal force distribution of tunnel structure [1]. However, the thickness of the segment determines the section stiffness of the tunnel structure. Under the condition of ensuring the normal clearance of the tunnel, properly increasing the segment thickness can effectively improve the overall transverse bending stiffness of the tunnel structure, enhance the deformation resistance of the tunnel structure, and further improve the service performance of the shield tunnel.Therefore, segment thickness is one of the important parameters for shield tunnel structural design [2].

At present, most researches have analyzed the difference of deformation and internal force variation of different segment thickness under design load through theoretical or numerical calculation. For example, document [3] takes an interval tunnel in Shanghai as an example, uses homogeneous ring method to check the structural internal force of shield tunnel, and points out that the increase of segment thickness can improve the stiffness of tunnel segment itself, but too large segment thickness is not conducive to the stress of tunnel structure itself. Document [4] adopts a calculation method that can consider the joint position and stiffness, analyzes the internal force changes under different segment thicknesses, and considers that simply increasing the segment thickness will lead to an increase in bending moment and concrete cracks. Document [5] calculated the influence law of segment thickness on internal force of large diameter tunnel under different surrounding rock levels by using modified conventional method, and gave the optimal segment thickness value. Document [6] compares and analyzes the deformation of pure steel structure segments and mixed segments under different thicknesses by ANSYS calculation.

However, subway construction is generally accompanied by surrounding development, which brings about large-scale adjacent engineering activities such as ground stacking [7-9], excavation and unloading [10-12], etc. When the external load changes too much, the tunnel structure will undergo severe convergence deformation, the internal force distribution of the structure will change greatly, and the segments will crack to varying degrees [13-16]. It can be seen that if the thickness of shield segments is designed according to the standard of design load, the applicability analysis of segments is not fully covered and the ultimate bearing performance of shield tunnel structure is reasonably evaluated.

Therefore, by establishing a refined three-dimensional numerical model of shield tunnel and verifying the model with full-scale tests, this paper analyzes the influence law of different segment thicknesses on the internal force of shield tunnel structure under overload and unloading conditions, with a view to providing reference for the structural design of shield tunnel.

2. 3-D Refined Numerical Model of Shield Tunnel

2.1 Segment Structure Type

The shield tunnel structure adopts an inner diameter of 5.5m and a ring width of 1.2m. The segment standard ring consists of six segments, including one capping block (K), two adjacent blocks (B1, B2), and three standard blocks (A1, A2, A3). The cap block corresponds to a central angle of 21.5, the adjacent block corresponds to a central angle of 68.0, and the standard block corresponds to a central angle of 67.5, as shown in Figure. 1. Segment rings are assembled in staggered joints according to the "A-B-A" pattern, and the positions of the adjacent ring capping blocks deviate from plus or minus 22.5 degrees directly above, as shown in Figure 2. The segment blocks are connected with each other by 2 hoop bolts, 12 hoop bolts are arranged for each hoop segment, and a longitudinal bolt is arranged between adjacent two hoop segments at intervals of 22.5 degrees, totaling 16 bolts. The longitudinal joints of the segments are provided with concave-convex tenons, while the circumferential joints are not provided with concave-convex tenons. Segment contains main reinforcement, longitudinal reinforcement, stirrup and bolt hand hole reinforcement. The diameter of main reinforcement and bolt hand hole reinforcement is 16mm, the diameter of longitudinal reinforcement is 10mm, and the diameter of stirrup is 6mm, as shown in Figure 3.

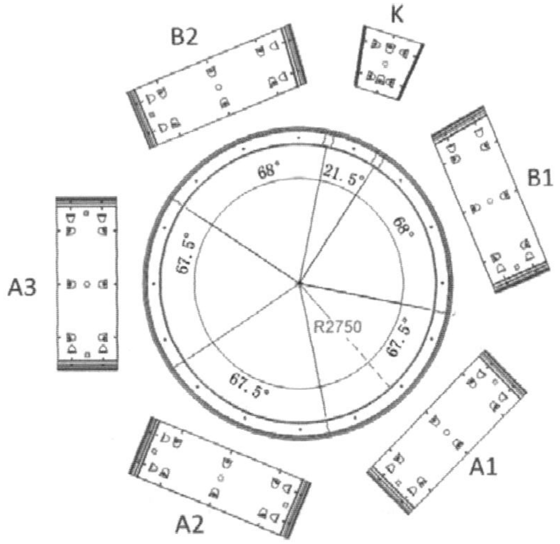

Figure 1. Schematic of the whole ring of shield tunnel

IOP Conf. Series: Materials Science and Engineering **894** (2020) 012018 doi:10.1088/1757-899X/894/1/012018

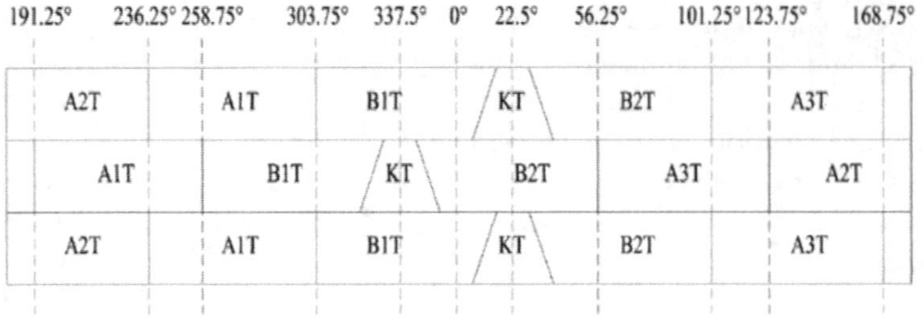

Figure 2. Shield segment assembling form

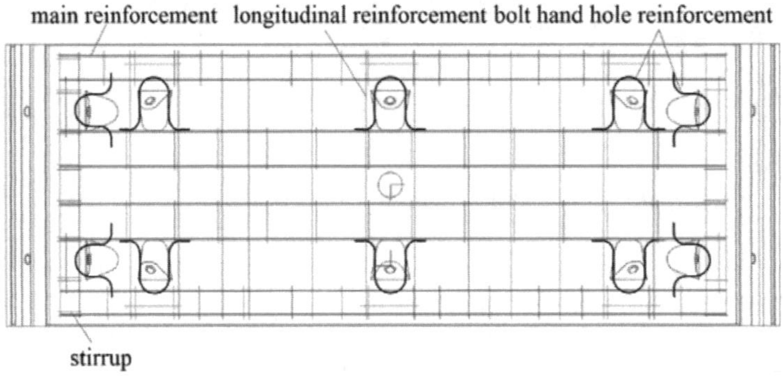

Figure 3. Schematic of standard block segment reinforcement

In order to analyze the influence of segment thickness on the internal force of shield tunnel under different loads (regardless of the change of waterproof gasket groove with segment thickness), three segments with different thicknesses were selected for modeling, with segment thicknesses of 350mm, 400mm and 450mm respectively, as shown in Figure 4.

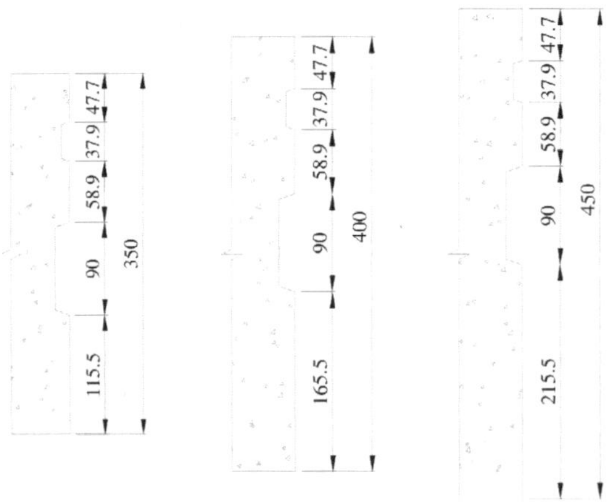

(a)350mm (b)400mm (c)450mm

Figure 4. Structural of different segment thickness

2.2 Segment Model and Material Parameters

Joint is the weak link of segment structure, and the accuracy of detail structure simulation is very important to truly reflect the mechanical properties of segment structure. Therefore, this paper fully considers the joint structure details such as longitudinal mortise, bolt hand hole, waterproof gasket groove, etc., and establishes a refined finite element model of three-ring segment by using large-scale general finite element software Abaqus.The single-ring segment finite element model is shown in Figure. 5, and the three-ring finite element model is shown in Figure 6.

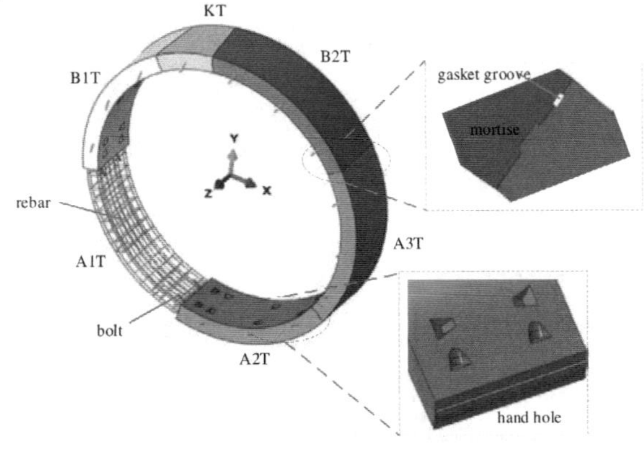

Figure 5. Refined finite element model of single ring segment

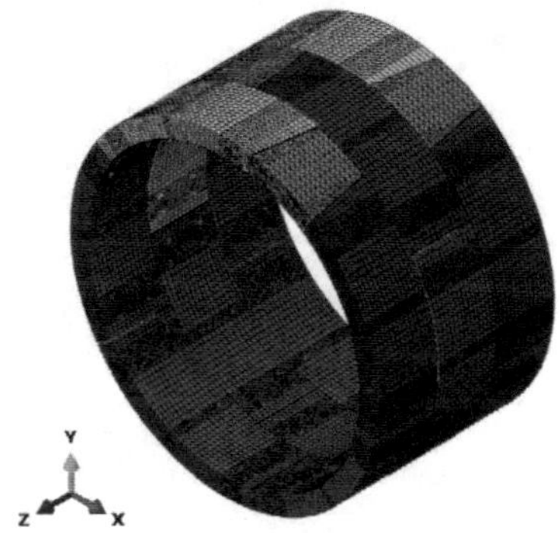

Figure 6. Three-ring finite element model

Segmental concrete is calculated using elastoplastic damage constitutive model [17]. The main reinforcement of the segment is HRB335 reinforcement, while the longitudinal reinforcement and stirrup are HPB235 reinforcement, ignoring the bolt hand hole reinforcement. The physical and mechanical parameters of other materials are shown in Table 1. The elastic modulus of the material after yield hardening is reduced to 0.01 times of the initial value, and its stress-strain relation curve is shown in Figure 7.

Table 1. Physical and mechanical parameters of bolts and reinforcement material

Material	Elastic modulus E/GPa	Poisson ratio μ	Yield strength f_y /MPa	Tensile strength f_u/MPa
5.8 grade bolts	200	0.3	400	500
HRB235steel bar	210	0.3	235	370
HRB335steel bar	200	0.3	335	455

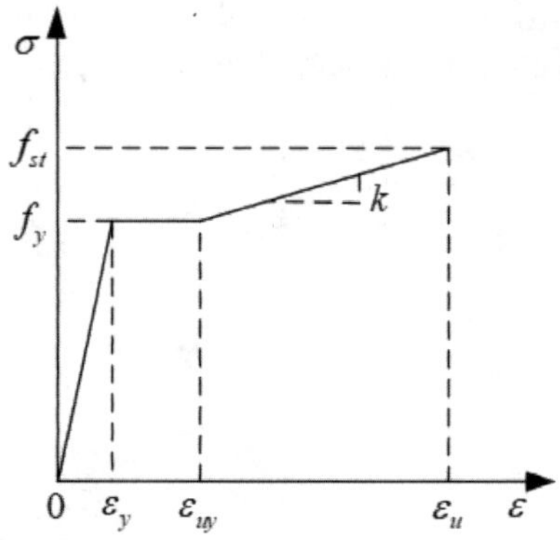

Figure 7. Stress-strain curve of reinforcement

Assuming that the stress-strain constitutive relation curves of bolts and steel bars under monotonic loading are consistent, as shown in Figure 8, the elastic-plastic characteristics of materials are characterized by three fold lines to simulate the yield, hardening and softening phenomena of bolts and steel bars during loading [18]. The expression of stress-strain relation in each stage is as follows:

$$\sigma_s = \begin{cases} E_s\varepsilon_s & \varepsilon_s \le \varepsilon_y \\ f_y & \varepsilon_y < \varepsilon_s \le \varepsilon_{uy} \\ f_y + k(\varepsilon_s - \varepsilon_{uy}) & \varepsilon_{uy} \le \varepsilon_s < \varepsilon_u \\ 0 & \varepsilon_s > \varepsilon_u \end{cases}$$

where: E_s is the elastic modulus, σ_s is the stress, ε_s is the strain, f_y representative value of yield strength, ε_y corresponding yield strain, ε_{uy} starting point of material hardening stage, ε_u corresponding peak strain and k is the slope of material hardening stage.

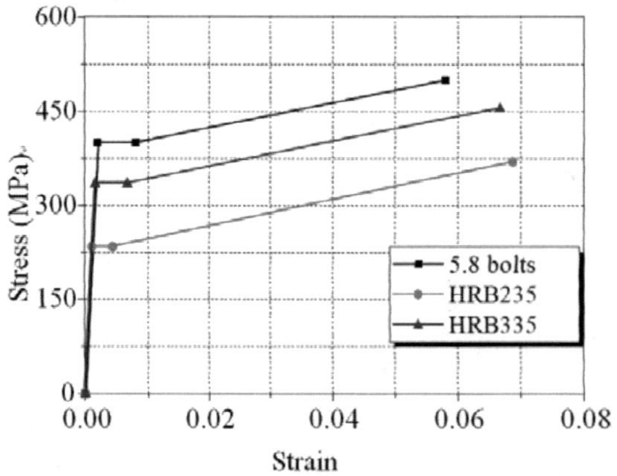

Figure 8. Stress-strain curve of shield tunnel reinforcement

2.3 External load conditions

The numerical calculation adopts the "load-structure" model [19]. as shown in Figure. 9, the top of the tunnel bears vertical water and soil pressure, the bottom bears foundation reaction P2, the horizontal loads P3, P4, P5 distributed symmetrically in trapezoid form on both sides are the tunnel dead weight, P6 is the lateral stratum resistance caused by tunnel deformation, and the loads are distributed uniformly along the longitudinal direction of the tunnel. The buried depth and load of the tunnel are shown in Figure. 10 and Table 2.

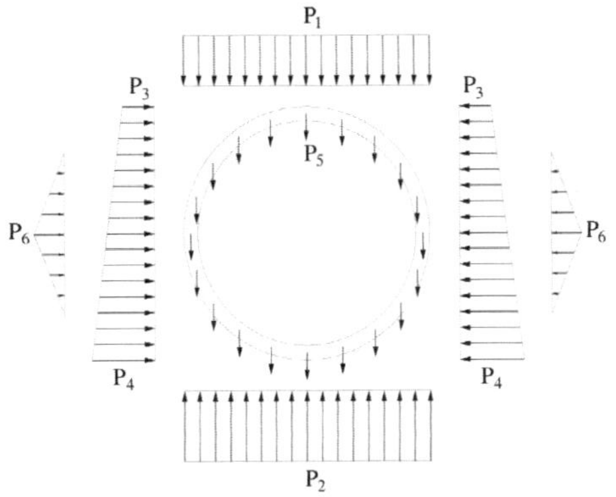

Figure 9. Load-Structure Model

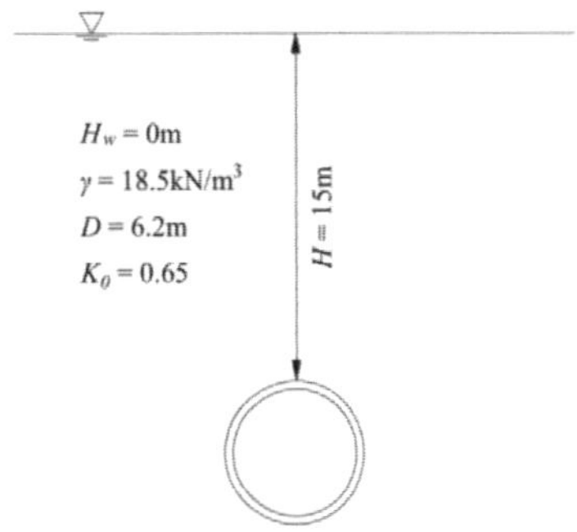

Figure 10. Schematic Diagram of Shield Tunnel Depth

Table 2. Load Values (kPa)

P_1	P_2	P_3	P_4	P_5
277.5	286.25	180.38	254.93	25.94

After the tunnel structure reaches the normal bearing condition, the overload condition is simulated by continuously increasing the vertical loads P1 and P2 and keeping P3 and P4 unchanged. By keeping P1 and P2 unchanged and reducing P3 and P4, the unloading situation is simulated.

3 Model validation

3.1 Full-scale Test Plan

In order to verify the correctness of the numerical model, a full-scale test load was carried out on the segment of the 3rd ring shield and compared with the numerical calculation structure.

In order to simulate the real stress state of the shield tunnel structure, 24-point concentrated symmetrical loading is adopted to simulate the actual external loads and dead weight of the shield tunnel structure, such as stratum load, soil resistance, surface loading and unloading, etc. The test loading device is shown in Figure 11.

Figure 11. Full-scale test loading device

In the experiment, 24 loading points were divided into 3 groups, P1, P2 and P3 respectively. the distribution of loading points is shown in Figure. 12. At the same time, in order to simulate the longitudinal restraint force between adjacent rings, 6 vertical loading points were used in the full-scale test to simulate the longitudinal restraint effect between segment rings. as shown in Figure. 13, the design load of each vertical loading point is 25t.

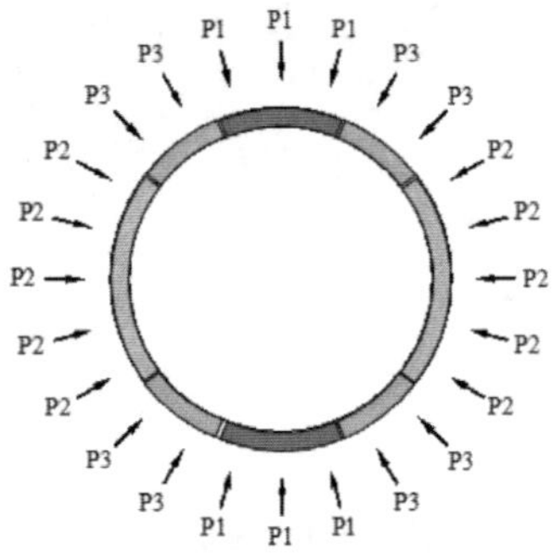

Figure 12. Test horizontal loading point

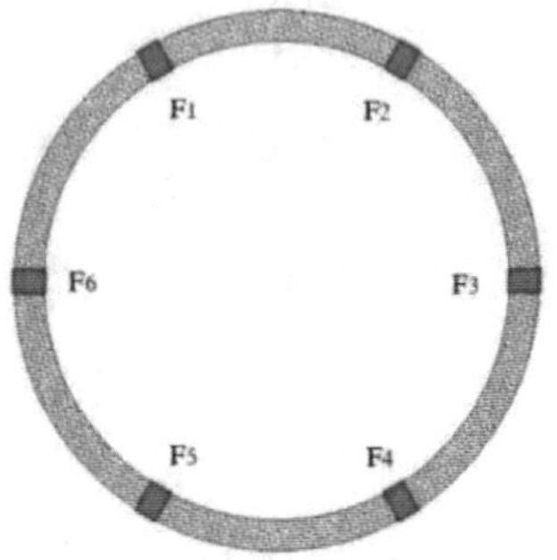

Figure 13. Test vertical loading point

The whole loading process of the test is divided into three stages:

(1) Load the vertical load to the design value, and P1 is graded from 0kN to F2. During the process, P2=0.7×P1 and P3=(P1+P2)/2=0.85×P1 are maintained, and the load increment of P1 at each stage is 0.1 times of F1.

(2) (2)P1 is loaded from F2 to F1. In this process, P2=0.7×P1 and P3=(P1+P2)/2=0.85×P1 are still maintained, and the load increment of P1 per stage is 0.05 times of F1.

(3) Continue to load P1 until the limit failure state. In this process, P2=0.7×P1, P3=(P1+ P2)/2, and the load increment of P1 is 0.05 times of F1.

Where: F1 is the design load and F2 is 0.8 times of the design load.

3.2 Comparative Analysis of Experiment and Numerical Calculation

Figures. 14 and 15 are comparative of convergence deformation and main reinforcement stress curves of full-scale test and numerical simulation. From the diagrams, it can be seen that the finite element calculation results are basically consistent with the full-scale test results, indicating that the model is reasonable and feasible.

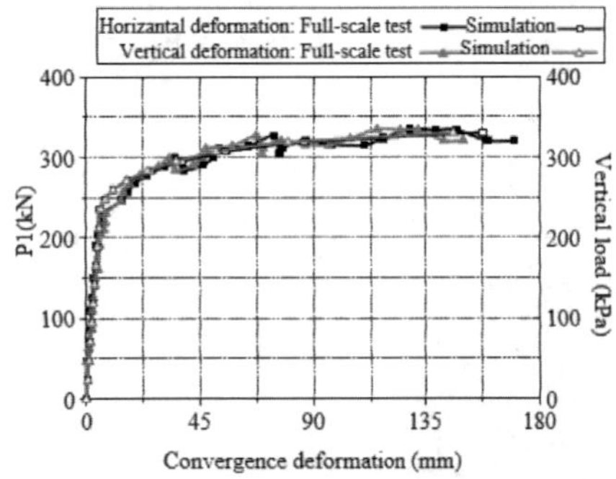

Figure 14. Horizontal convergence deformation comparison curve of test and numerical calculation

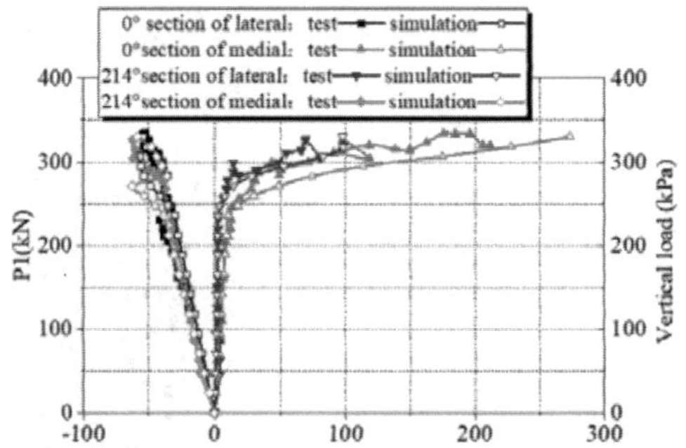

Figure 15. Stress comparison curve of main reinforcement between test and numerical calculation

4. Influence of Segment Thickness on Internal Force of Shield Tunnel

4.1 Variation Law of Bending Moment and Axial Force under Overload Condition

The variation rules of section bending moment and axial force of tunnel structures with different segment thicknesses with vertical load are shown in Figure. 16-21 respectively.

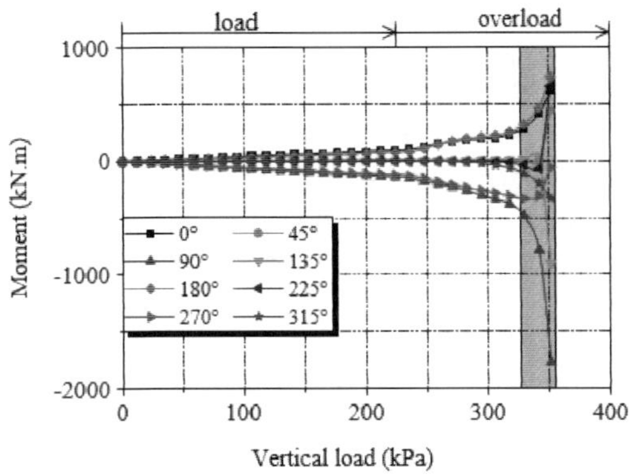

Figure 16. Variation of tunnel section bending moment when segment thickness is 350mm under overload

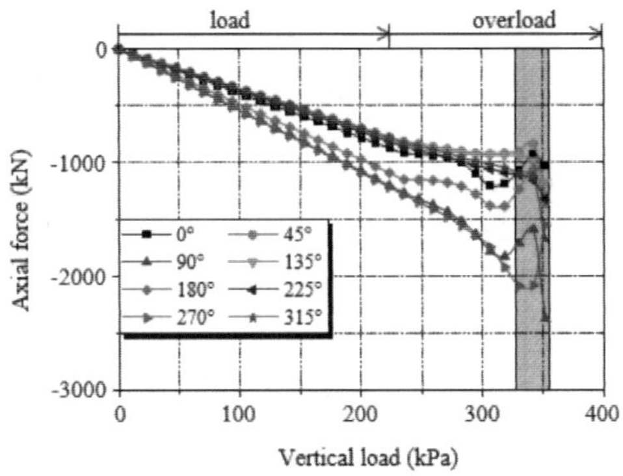

Figure 17. Variation of axial force of tunnel section when the thickness of segment is 350mm under overload

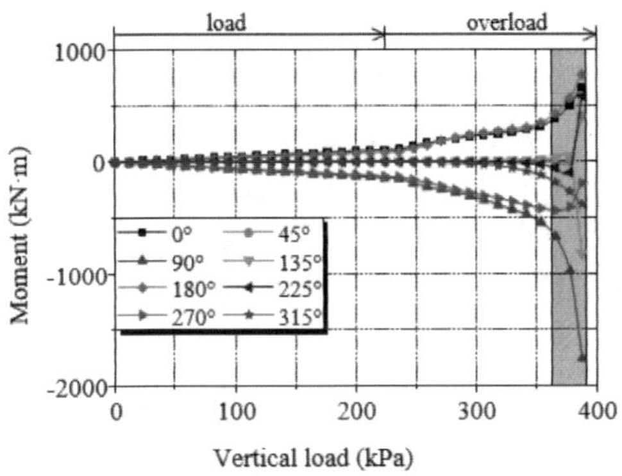

Figure 18. Variation of tunnel section bending moment when segment thickness is 400mm under overload

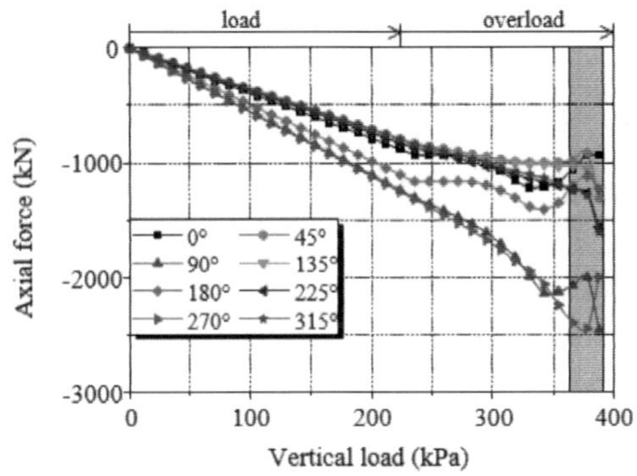

Figure 19. Variation of axial force of tunnel section when the thickness of segment is 400mm under overload

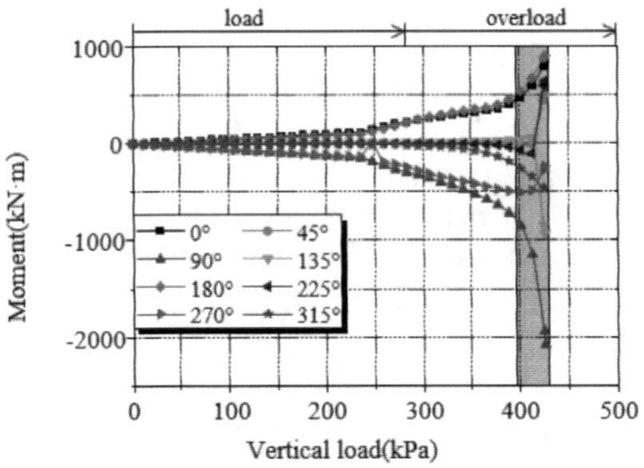

Figure 20. Variation of tunnel section bending moment when segment thickness is 450mm under overload

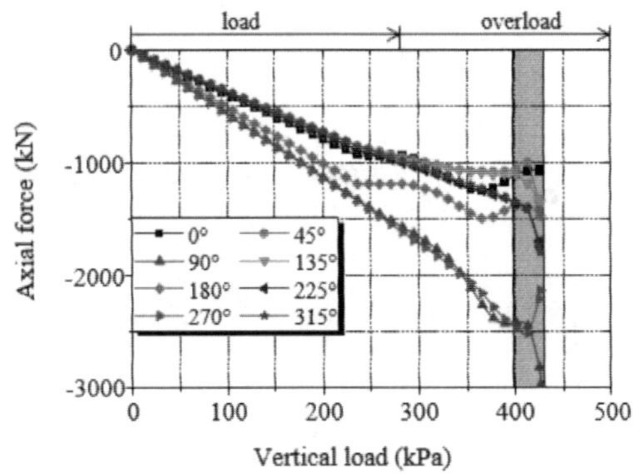

Figure 21. Variation of axial force of tunnel section when the thickness of segment is 450mm under overload

As can be seen from Figures. 16-21, in the normal bearing stage, the bending moment and axial force of each section of segments with different thicknesses increase linearly with the increase of vertical load. After entering the overload stage, the tunnel structure gradually enters the elastoplastic state, the section bending moment increases and the trend is faster and faster, and the fluctuation of section axial force increases. When the vertical load reaches 330.4kPa (350mm thick), 365.8kPa (400mm thick) and 401.2kPa (450mm thick), 1.57 times, 1.74 times and 1.91 times of the designed vertical load respectively, it can be seen that the bending moments of 0, 45, 90, 135, 180, 225 and 315 sections increase sharply while the bending moments of 270 sections decrease instead. The axial forces of sections 0, 90 and 180 increase to form reverse bending points and then decrease rapidly, while the axial forces of sections 45, 135, 225 and 315 fluctuate and then increase rapidly, which indicates that the tunnel structure reaches the critical point of instability failure. This state is defined as "critical instability state", the left boundary of the shaded part in the Figure indicates that the tunnel structure is in "critical instability state", while the right boundary of the shaded part indicates that the tunnel structure is in "ultimate failure state". Therefore, the larger the thickness of the segment, the larger the corresponding critical buckling load, and the inflection point of bending moment and axial force will lag.

The distribution of bending moment and axial force of tunnel structures with different segment thicknesses under different loads are shown in Figures. 22 and 23 respectively.

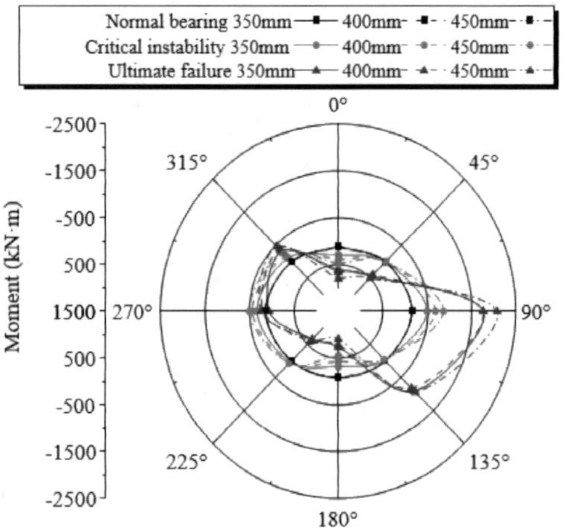

Figure 22. Bending moment distribution curves of tunnel structures with different segment thicknesses under overload

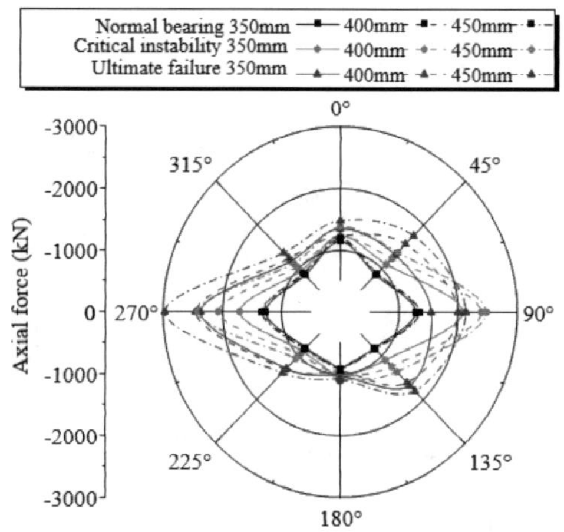

Figure 23. Axial force distribution curves of tunnel structures with different segment thicknesses under overload

As can be seen from Figures. 22 and 23, the bending moment and axial force of the cross section increase with the increase of the thickness of the segments, and the structural performance of the tunnel deteriorates with the increase of overload, and the difference in internal force between different segment thicknesses becomes more obvious.With the increase of load, the positions of maximum positive and negative bending moments of shield tunnel structure will change, but the positions of maximum positive and negative bending moments of tunnels with different segment thicknesses under the same state will not be affected.

4.2 Variation Law of Bending Moment and Axial Force under Unloading Condition

The variation law of cross-sectional internal force of different segment thicknesses under unloading condition with horizontal load is shown in Figure. 24-29 respectively. From the Figure we can see that the horizontal load in abscissa refers to the horizontal stratum load borne by the waist position of the tunnel.

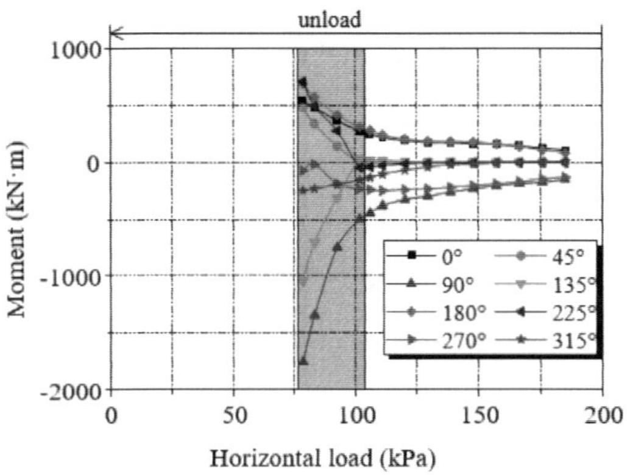

Figure 24. Bending moment variation of tunnel section when segment thickness is 350mm under unload

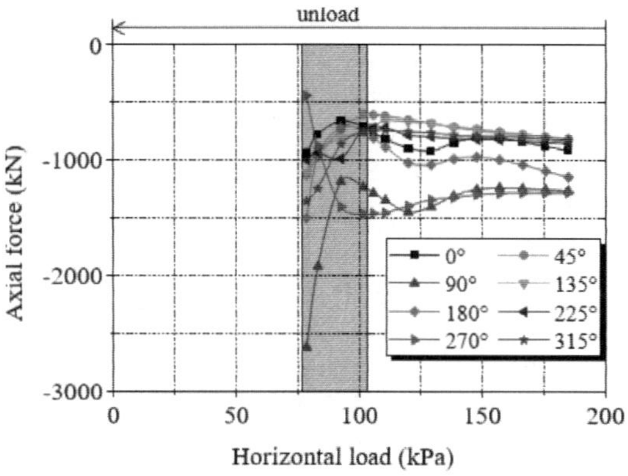

Figure 25. Axial force variation of tunnel section when segment thickness is 350mm under unload

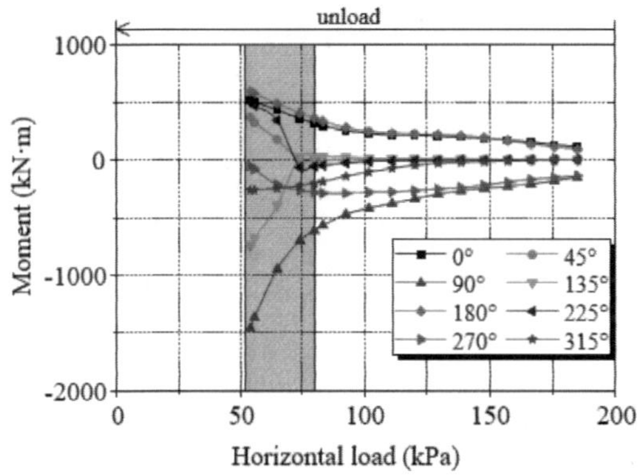

Figure 26. Bending moment variation of tunnel section when segment thickness is 400mm under unload

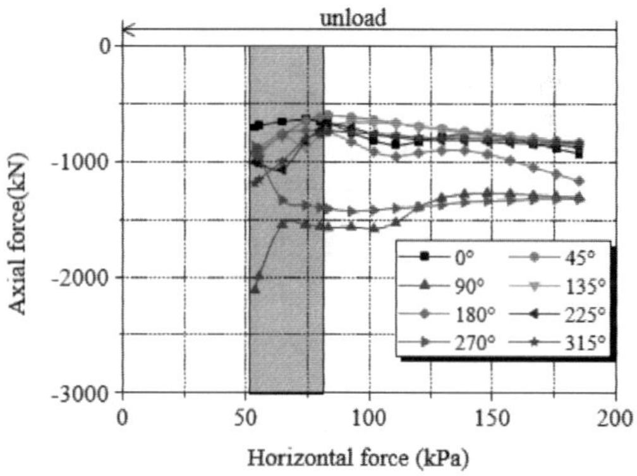

Figure 27. Axial force variation of tunnel section when segment thickness is 400mm under unload

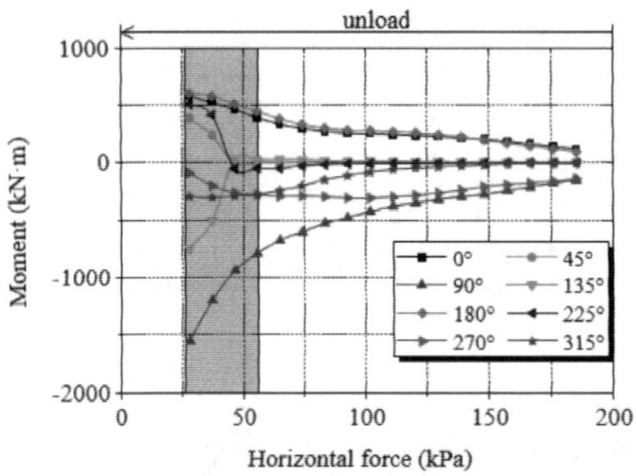

Figure 28. Bending moment variation of tunnel section when segment thickness is 450mm under unload

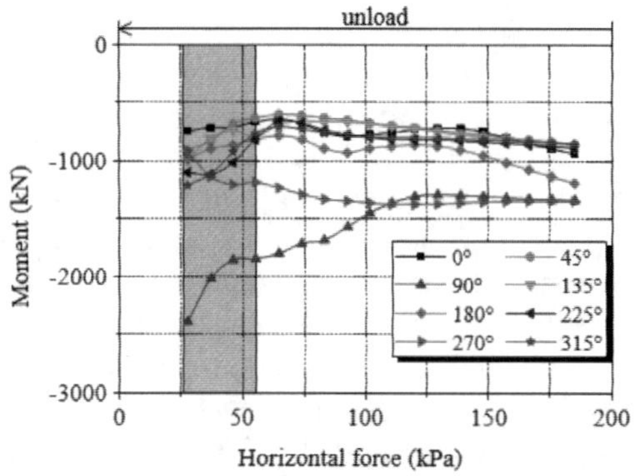

Figure 29. Axial force variation of tunnel section when segment thickness is 450mm under unload

As can be seen from Figures. 24-29, with the increase of unloading degree, the horizontal load decreases continuously, the section bending moment increases continuously. Except that the axial force of 90° and 270° sections increases with the decrease of the horizontal load, the axial force of other sections decreases with the decrease of the horizontal load, and the decreasing trend becomes slower with the increase of the thickness of the segment. When the horizontal load decreases from 185.0kPa to 105.9kPa (350mm thick), 79.8kPa (400mm thick) and 55.5kPa (450mm thick), respectively, the section bending moment begins to increase rapidly, and the axial force curve of some sections turns and begins to increase reversely, indicating that the tunnel structure reaches the "critical instability state" and begins to undergo instability failure. After that, when the horizontal load continues to decrease to 78.6kPa, 53.8kPa and 27.7kPa, the tunnel structure reaches the "ultimate failure state".

Figures. 30 and 31 are sectional bending moment and axial force distribution diagrams of tunnel structures with different segment thicknesses under different loading conditions under unloading conditions respectively. It can be seen that under the unloading condition of "critical instability state", the section bending moment and axial force also increase with the increase of segment thickness, and the difference between bending moment and axial force becomes more obvious with the further increase of unloading level. In addition, the bending moment and axial force near the 90 section position in the "critical instability state" are much larger than those in other sections.

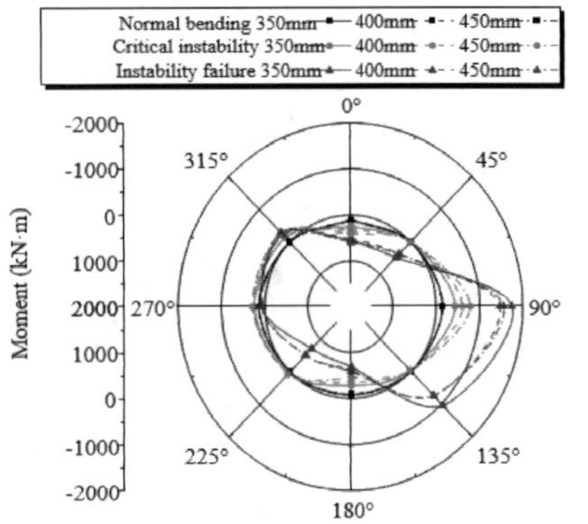

Figure 30. Bending moment distribution curves of tunnel structures with different segment thicknesses under unload

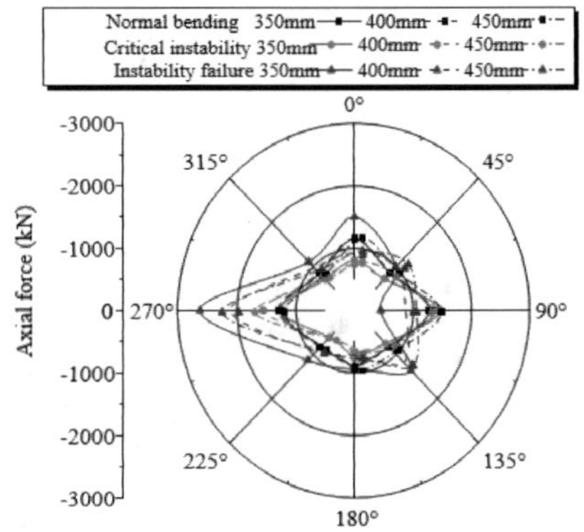

Figure 31. Axial force distribution curves of tunnel structures with different segment thicknesses under unload

5. Conclusions

In this paper, by establishing a three-dimensional refined numerical model of shield tunnel, the influence law of different segment thickness on the internal force of shield tunnel structure under overload and unloading conditions is studied, and the following conclusions are obtained:

(1) In the process of overload and unloading, the internal force of each section of tunnel structure increases with the increase of segment thickness, and the internal force difference among different segment thickness increases with the aggravation of overload and unloading degree.

(2) The increase of segment thickness will make the sudden change point of bending moment and axial force (namely critical instability state) of shield tunnel lag.

(3) The increase of bending moment and axial force in the process of segment thickness increasing from 400mm to 450mm is obviously larger than that in the process of segment thickness increasing from 350mm to 400mm, which indicates that excessive increase of segment thickness will greatly increase the internal force of the tunnel structure, resulting in the decrease of crack resistance of the tunnel structure.

6. References

[1] Huang Z H, Liao S M 2000 Effect of Segment Thickness on Mechanical and Service Performance of Tunnel. *Architectural Technology,* **31**(7):471-472.

IOP Conf. Series: Materials Science and Engineering **894** (2020) 012018 doi:10.1088/1757-899X/894/1/012018

[2] Zhu S Y 2002 Current Situation and Development of Segment Structure Design for Shield Tunnels in China. *Modern Tunnel Technology,* **39**(6):23-27.

[3] Huang Z H, Liao S M, Liu G B 2000 Optimization of Segment Thickness of Shanghai Soft Soil Shield Tunnel. *Geotechnical Mechanics,* **21**(4):326-330.

[4] Zhao G X, He C 2003 Optimization Analysis of Tunnel Segment Design [J]. *China Railway Science,* **24** (6): 61-66.

[5] Chen W T, Fu Z Q, Ma J X 2017 Study on Segment Thickness Design of Large Diameter Double Shield TBM Tunnel. *Railway Architecture,* **57**(10):60-62.

[6] Zhao Z P, Liu G B, Bai Y 2018 Segment Design of Large Diameter Shield Tunnel with High Water Pressure. *Highway Traffic Technology,* **34**(S01):102-110.

[7] Huang D W, Zhou S H, Lai G Q 2017 Degradation Mechanism and Characteristics of Shield Tunnels under Surface Overload. *Journal of Geotechnical Engineering,* **39** (7): 1173-1181.

[8] Zhang M G, Zhou S H, Huang D W 2016 Analysis of Influence of Surface Overload on Subway Shield Tunnel. *Geotechnical Mechanics,***37** (8): 2271-2278..

[9] Wang R L, Zhang D M 2013 Study on Lateral Deformation Mechanism and Control Index of Shield Tunnels in Soft Soil under Overload. *Journal of Geotechnical Engineering,* **35**(6):1092-1101.

[10] Dai Z R 2017 Study on Fragmentation Mechanism of Subjacent Shield Tunnel Segment Caused by Large-scale Excavation and Unloading on Surface. *China Railway Science,* **38** (4): 62-69.

[11] Shen G G, Ding J F 2016 Measurement and Analysis of Impact of Loading and Unloading on Existing Shield Tunnels. *Journal of Nanjing Institute of Engineering* , **14**(04):61-64.

[12] Yao A J, Zhang J T, Guo H F 2018 Study on Unloading-Loading Effect of Foundation Pit Excavation above Subway Shield Tunnel. *Geotechnical Mechanics,* **39**(7):2318-2335.

[13] Liu C, Lai J X, Liu Q 2013 A review of research situation on shield tunnel diseases. *Applied Mechanics and Materials,* **6**:438-439.

[14] Dong F, Fang Q, Zhang D L 2017 analysis of disease state of Beijing subway tunnel. *Journal of civil engineering,* **50**(06):108-117.

[15] Yang Y B, Zhou B, Xie X Y 2016 Study on Transverse Deformation and Cracking Characteristics of Shield Tunnels under Adjacent Foundation Pit Construction. *Journal of Rock Mechanics and Engineering,* **35**(S02):4082-4093.

[16] Yang X, Gao Y 2012 Discussion on Causes of Segment Cracking in Shield Tunnels during Subway Operation. *Urban Rail Transit Research,* **15**(7):26-29.

[17] Jin H, Yu S, Zhou S H 2019 Research on mechanics of longitudinal joint in shield tunnel by the nonlinear spring equivalent method. *KSCE Journal of Civil Engineering,* **23**(2):902–913.

[18] Jin H, Yu K W, Gong Q M 2018 Load-carrying capability of shield tunnel damaged by shield shell squeezing action during construction. *Thin-Walled Structures*, **132**:69-78.

[19] Yu S, Jin H, Zhou S H 2019 Morphology of Rust Layer in Shield Segment Reinforced by Chloride Ion and Miscurrent. *Engineering Mechanics,* **36**(7):174-183.

AUTHOR INDEX